Quantum Meaning

Quantum Meaning

A Semantic Interpretation of
Quantum Theory

Ashish Dalela

SHABDA
PRESS

Quantum Meaning—A Semantic Interpretation of Quantum Theory
by Ashish Dalela
www.shabda.co

Published by Shabda Press
www.press.shabda.co
ISBN 978-81930523-7-2
Second Edition
v1.5(06/2021)

If quantum mechanics hasn't profoundly shocked you, you haven't understood it yet.

—Niels Bohr

Contents

List of Figures

Preface

Quantum theory presents such an unintuitive picture of reality that it has now become customary to supplant the mathematics in the theory with a philosophical interpretation about its real meaning. No other scientific theory has had such a wide variety of interpretations as quantum theory, which tells us that the theory is not as easily understood as other theories. And yet, despite many interpretations, we aren't any closer to comprehending the real meaning of quantum theory today than we were a hundred years ago when interpretations began. Interpretations underscore our need to *understand* what is really out there rather than what is just being described by the theory. This human need to go beyond what is empirically proven and can be said unambiguously through mathematics can prove frustrating for some physicists. They would rather insist that we "shut up and calculate," as Feynman once supposedly said.

At the root of this stance is the idea that there is no scientific merit in interpreting the theory because the theory is final. Rather than try to understand what it means in terms of other concepts, we might as well get used to the quantum concepts. The scientific need for interpretations thus boils down to just one question: Is the present quantum theory a final theory of reality? If it is, then the sooner we get used to quantum ideas, the better it is for everyone. If it isn't, then the engaging debates about the nature of quantum principles are indeed useful stepping stones to a better theory.

I believe that current quantum theory is not a final theory for the following three reasons. First, the application of quantum principles to macroscopic objects leads to a variety of problems, including the famous measurement problem. Quantum theory applies even to the macroscopic world although we tend to selectively apply it only to atomic objects. Without solving the measurement problem, and giving a quantum explanation to the macroscopic world, we cannot call quantum theory a complete theory. Second, we need a clear conception about the nature of atomic objects which leads to unintuitive (relative to classical physics) ideas such as non-locality, wave-particle duality, and the uncertainty principle. Without understanding the nature of these particles, and why they behave differently from classical particles, we cannot make progress in quantum theory. Third, based on this understanding of quantum particles, we need to formulate an understanding of change. The classical idea about change as motion has collapsed in quantum theory, although a new picture about what we mean by change has not emerged. The classical picture of motion assumed immutable particles that changed state without changing identity. This is no longer sustainable in quantum theory because the state and the identity cannot be distinguished; a new state is now a new particle. Particles don't move in quantum theory; they are transformed into new particles. A new notion of change is needed to explain this.

To solve these problems, we need to provide an intuitive picture of reality for both macroscopic and atomic phenomena. This picture must explain change, but in a way different from the classical notion of motion. It must also explain the reason why quantum particles behave non-locally, exhibit wave-particle duality, and their state is governed by uncertainty relations. I believe that a theory that fulfills these objectives will have to discard many assumptions made in classical physics. If quantum theory is such a theory, then it cannot be an incremental progress on classical physics but a fundamentally different type of theory. Everyone recognizes that quantum theory is different from classical physics although the task of interpretation dwells only upon the gaps vis-à-vis classical physics, while retaining most classical dynamical concepts and notions about causal laws. My approach to interpretation differs in this respect because I believe that quantum theory provides a view of reality radically different from classical physics. This view of reality is indicated by quantum problems but not captured by the current interpretations.

This new view of reality must be drawn from a different part of everyday experience than which science has considered so far. I am referring to the phenomenon of meaning. Not the fact that science describes knowledge of reality but that we can represent knowledge in matter. These representational capabilities are exhibited when we treat the material world as symbols of meaning. To represent knowledge, matter must have properties that allow us to embed information about reality within reality. Classical physics studies this information in terms of material properties such as mass, charge, momentum and energy which is like studying a book in terms of its weight or speed. The same world can be described as symbols instead of physical properties in which meaning will gain prominence over the physical form of the symbol—e.g. size, weight, shape, etc.

This radically changes the nature of causality. For instance, changes in objects will not be about the motion of particles but about the evolution of knowledge. Motion is continuous, but the evolution of knowledge is discontinuous. Classical particles are dimensionless points, but symbols have a finite size. Physical properties can be defined in relation to universal measuring standards, but meanings are defined always in relation to other symbols in a collection. Classical particles are independent of each other, but meanings are defined in opposition to other meanings, or not at all. For instance, it is impossible to define 'hot' without defining 'cold'. These two meanings are mutually 'entangled' even though separate.

My interpretation of quantum theory starts from the idea that we have made a mistake in classical physics by treating the world as things rather than symbols. Things we currently consider unintuitive or bizarre about quantum theory can be addressed by revising our view of reality from particles to symbols. An interpretation that explains the current set of problems in terms of unaccounted features in reality presents a distinct scientific advantage over interpretations that view it as a final theory. If matter can hold meanings, then matter has additional properties that have not been conceived in science so far. This work argues that the meaning of quantum theory is that it is a theory about meaning. The bizarre aspects of the theory arise because we describe symbols in terms of classical physics which deals with objects without meaning. The bizarreness in quantum theory is then only apparent, and can be resolved when matter is understood in terms of its *representational* and *computational* properties. This requires a rethink of quantum principles. The present work is thus entitled "Quantum Meaning" rather than "The Meaning of Quantum Theory." I believe that the former is a scientific problem while the latter is merely a philosophical problem.

The difference between a physical property and meaning is that between *quantity* and *type*. The reinterpretation of physical properties as meanings requires objects to be seen as *denoting types* rather than *having quantities*. A sign such as "$" has a shape, but it denotes a type associated with a meaning. The observable remains the same in the two cases, but the observable is perceived in two ways—percept and concept—rather than just one (percept). Indeed, even the percepts must now be defined in terms of concepts (e.g. color in terms of concepts like 'red', 'yellow', and 'blue') rather than quantities (such as the frequency of a wave). Both percepts by which we detect symbols, and the meaning associated with those symbols, therefore, become concepts. The implication of this new way of thinking is that causality is attributed to concepts rather than quantities. These concepts include both sense-observable percepts but also the mentally perceivable meanings. Semantics introduces new forms of causality quite different from classical causality.

For instance, semantically it is possible to explain why people stop at a red light even though a theory based on the frequency of light itself cannot explain this fact. Reinterpreting physical states as meaning doesn't change observables although it changes the *causal explanation* of behaviors. By changing causality, many things that we cannot explain using classical forms of causality can now be explained. The semantic view therefore ushers in a new type of causal explanation that is not possible in current physics. A semantic view of quantum reality also helps us understand why a classical treatment of semantic states will suffer from indeterminism, probability, uncertainty, wave-particle duality and non-locality. For instance, the redness of light doesn't always mean stop. In some cases, it can mean danger, which would imply running away. This illustrates the idea that a theory that reduces semantic states to physical states will be incomplete because it will not be able to explain whether redness denotes stopping or running away.

The formal mathematical theory of semantic information is not developed here, although a sketch of a formalism where meaning is represented via extended forms within space is described. This requires us to view spatial extension—length and position—as semantic rather than physical properties. The present interpretation sets up insights and intuitions about a newer way of thinking about space, which can lead to a new mathematical description of quantum phenomena. The development of a full-fledged mathematical theory about meanings in symbols, however, requires a solution to the problems of semantics within mathematics, which are not addressed here. I have, however, discussed these in *Gödel's Mistake*, which describes paradoxes in logic and mathematics and connects these paradoxes to the nature of meanings in ordinary language.

If the quantum world is a world of symbols, atomicity is the limit to the divisibility of information in material symbols. Information must come to us in chunks; the smallest bit of data is the most elementary idea and the smallest algorithm is the most elementary operation. Chunks of information are *bits* that make up the universe. We can liken these bits to letters that form propositions. In current physics, physical properties are causally relevant but semantic properties are not. Quantum theory can be completed if semantic properties have causal effects. This is like how we decode the information in a book not in terms of its weight and length, but in terms of a *language* in which physical tokens are viewed as symbols of meaning. Explanatory gaps can be bridged if the theory accounts for what quanta *represent* in addition to what they *are*.

The Semantic Interpretation of Quantum Theory is the view that meaning is part of the universe but not described by any fundamental theory of physics. Meaning is not an epiphenomenal property of macroscopic objects—brains, computers, etc. It is rather a property of the fundamental atomic objects themselves. If quanta are capable of being symbols, then nature has fundamental representational and computational properties, unknown to physical theories so far. Postulating the existence of meaning to complete quantum theory does not make it a hidden variable theory because meanings are derived from the same empirical facts as current physics. Semantics rather helps us understand why a non-semantic theory must have probabilities, uncertainty, non-locality and wave-particle duality.

My goal therefore is not to *interpret* the current quantum formalism in a way that rationalizes its problems—as is the case with the current interpretations. My goal is rather to provide an interpretation that shows why the quantum bizarreness exists so that we can conceive an alternative theory of nature based on meaning. The general purpose of interpretations is to take the symbols of a theory for granted and provide a meaning to them. My interpretation will rather *not* take the symbols for granted; I will rather aim to show that this theory is not just incomplete, but it is *wrong* in the sense that it carries forward classical assumptions in which matter and mind were separated as two 'substances'. My claim is that there aren't two substances to contend with. There is just one type of reality, but it is symbolic; so, it has a physical existence in the sense that we can observe it, but it also has meaning. The physical existence is also described in terms of types rather than quantities. So, we are dealing with a reality that comprises many types, and the causality is no longer in the quantities but in the types. Once we redefine the nature of causality, we can redefine the natural laws. Current quantum theory is therefore a vista to a revolution. The revolution involves an overhaul of thinking about the world.

To those who consider quantum theory a revolution over classical physics, I can only say: the revolution is incomplete. We have taken the classical physical concepts and defined new mathematical relations between them, rather than adopt a new set of concepts. It was important for the founders of quantum theory to maintain a sense of *continuity* with classical physics, which is unnecessary. This need arose from the presumption that the *macroscopic* world is described adequately by classical physics, and we need a new theory of the *microscopic*. I would instead argue that even the macroscopic objects—e.g. books, painting, musical compositions, or drama—are incompletely described when we view them as classical physics. We don't need to maintain continuity with classical physics; we rather need to reject it completely in favor of a wholesome theory that doesn't distinguish between the nature of big and small.

A physical theory cannot be a victim of sizes—which are relative to our perceptual apparatus. Who is to say, for example, that something is small or big? It completely depends on our senses. If we had the senses that could perceive the microscopic, then it would be macroscopic. So, to pretend that we need a separate theory for the atomic world is to assume that nature's theories are dependent on our perceptual apparatus, something we should aim not to. Thus, the belief that classical physics is adequate for the macroscopic world is not only inadequate when we look at macroscopic symbols, but it is also misleading in making us think that atomic theory is about the microscopic. These mistakes can be corrected by breaking away from the classical physical notion about matter and its properties.

1

Quantum Information

How wonderful that we have met with a paradox. Now we have some hope of making progress.

— *Niels Bohr*

Setting the Scene

Quantum theory is now more than a century old. When, in 1900, Max Planck presented his landmark paper on black body radiation, which introduced the idea of a quantum of energy, most prominent physicists did not believe in the existence of atoms let alone sub-atomic particles. Matter was supposed to be this infinitely divisible and continuous substance whose properties got smaller and smaller in magnitude as we divided it into more and more granular pieces. So, the initial idea that matter may not be in fact infinitely divisible was itself a major revolution in the physics community. But today, most of modern technology including communications, electronics, space exploration and biotechnology depend on it. Quantum theory is the most successful and valuable theory science has ever built.

Notwithstanding its many successes, the fundamentals of quantum theory are still quite mysterious. In fact, they are as mysterious today as they were when the theory was originally formulated. Early founders of the theory including Planck, Einstein, Bohr, Born, Heisenberg and Schrödinger debated the meaning of quantum tenets and their exact significance. It was important for them to *understand* the theory and its wider philosophical implications. But in the last sixty years, physics teachers like Richard Feynman and Freeman Dyson have replaced the

1

quest for meaning with the computation of numbers. Quantum theory, in this view, does not need an interpretation if it can make correct predictions. Although there are occasional papers interpreting the theory's meaning, writing about the philosophy of quantum theory isn't the most respectable professional choice for physicists. It is still a respectable professional choice for philosophers, although what a philosopher says does not make headlines in the physicist's world.

Physicists widely believe that speculating upon the meaning of quantum theory is unnecessary. It is only necessary to understand the rules by which the game of quantum theory must be played. After all, we wouldn't waste time trying to understand why a football field has a certain size, or why a tennis ball must be of a certain diameter. The rules of quantum theory are no more meaningful than the rules of other games. The quantum formalism is a mathematical theory that makes predictions. The methods by which we compute these measurable outcomes don't need to have ulterior meanings.

Underlying this widespread view is the belief that an interpretation of a theory does not add value to the theory itself. There are many ways you can look at a theory, which are all consistent with the mathematics. These views are basically different *models* of the theory. As an example, suppose that you were asked to interpret the equation 'A + B = C.' Depending upon how you look at it, this equation has either no meaning or several meanings. It can represent the outcome of adding 5 apples and 5 oranges to get 10 fruits. It can also represent the total number of people in the world if the count of people below the age of 50 is A and the count of people above 50 is B. Countless other interpretations of this formula can be created because A and B can be made to refer to many different types of entities. The act of interpreting creates one particular way of understanding the formula—sometimes called a *model* of the theory—although it is not necessary to have models. Physicists similarly believe that interpretations of quantum theory will not change the theory itself and each interpretation is one type of model by which we could *understand* the theory although no one view will eliminate other ways of understanding. Critics of interpretations thus believe that the mathematical theory is compatible with all interpretations, and an interpretation is our view of how we want to think about the theory, without impacting the theory itself.

What I will present in this book significantly differs in this respect. I will describe an interpretation that, like any other interpretation does due diligence of explaining the meaning of uncertainty, probability, non-locality, quanta, and other concepts in quantum theory. But my goal is not to create yet another model of quantum theory. My goal is rather to create a model of reality, which can be used to understand the shortcomings in the present theory. If reality is different than how we have conceived it in current physics, then an alternate picture of reality can help explain why that reality would be inadequately described by current quantum theory. The alternate picture can also be used to develop an alternate mathematical theory of nature that explains the observations in a new way. The alternate theory will not be probabilistic and causally incomplete. Rather, the new theory will produce predictions which current theory cannot because of its unavoidable probabilities.

The interpretation of quantum theory described in this work is called the *Semantic Interpretation of Quantum Theory* (SI henceforth). SI has both philosophical and scientific implications. The key difference in SI vis-à-vis other interpretations of quantum theory is that probability, wave-particle duality, uncertainty and other problematic features of quantum theory are shown as limitations of the theory, not of reality. Quantum objects are not just units of matter but also units of information. Probability, uncertainty, wave-particle duality, etc., which make quantum theory mysterious vis-à-vis classical physics, are problems that arise when we think of information in terms of classical physical concepts. SI discusses how the mysteries disappear if reality includes information.

The Quantum-Classical Conflict

The biggest problem bedeviling our understanding of quantum phenomena is our attempt to think of quantum reality in terms of classical concepts. Since classical concepts are inadequate for describing quantum phenomena, notions such as uncertainty, probability, non-locality and complementarity have been added as conceptual addendums to the classical view. The task of interpreting the theory now expends efforts in trying to explain the addendums, given that we seem to already understand the classical worldview. Early quantum theorists were bred on classical physics and wanted to broker peace between the established classical view and the new quantum principles. A path of least resistance was thus chosen where we reuse ideas from classical physics, making quantum theory an incremental progress on classical physics. Specifically, it was very important for the early founders to show that quantum theory *reduces* to classical physics under some conditions. It was also important to retain classical measurements, as it seemed that these are the only way we can empirically know the world. Early quantum theorists were convinced that the macroscopic world is classical.

While the incremental approach to explaining quantum phenomena did create a predictive framework, it also brought two problems. First, by reusing the language of classical physics, quantum theory retained the classical intuitions although their meanings were inapplicable. Second, to reconcile the disparity between classical and quantum formalisms, the theory added many unintuitive ideas making it very hard to understand. The multiple interpretations of the theory in the last century haven't made our understanding of quantum phenomena any better, because these essentially try to create a *model* of the quantum theory without actually addressing the issues about uncertainty, non-locality, probability and discreteness that started the entire debate. Can we now go back in time and think about quantum principles without the prejudices of classical mechanics constraining our philosophy about nature?

A rethink of quantum principles needs to ask a very fundamental question: Is the everyday world a classical world? Despite the many successes of classical physics, I do not believe that the everyday world is classical because it contains information—pictures, books, music and science—which cannot be described by classical physics. The phenomena of information are widespread in nature, and current physical theories neglect these phenomena. Classical physics studies matter as if it were devoid of information, and this works when information can apparently be reduced to classical properties. But there are also instances when a reduction of macroscopic objects to classical properties is not possible. In these cases, classical physics is bound to fail, even at the level of macroscopic objects.

The genesis of the informational problem is that classical physics describes objects in a context-independent manner whereas objects acquire meaning in relation to other objects. Some of these meanings may refer to the properties of other objects. Pictures, books, music and science are, for instance, *about* other things. The properties of a classical object depend only on that object, and the object describes nothing other than itself. The meanings of a symbol on the other hand depend on other objects and a symbol can describe other objects or even other symbols. These two basic differences make classical physics an inappropriate theory for symbols.

A universe of classical particles is devoid of *knowledge* because the universe can only be itself and not be a representation. If the universe were only composed of classical particles, then there would only be physical properties, but no meanings. The idea that we can have information about an object without becoming that object is central to all knowledge acquisition. By holding information about the world, we create its representations and interpretations. There is now a distinction between what we know about the world and what the world is in itself; specifically, knowledge can be false. Thus far, there is nothing in science that permits the creation of knowledge about other objects because fundamentally there cannot be things that are semantically false although they exist physically. In a sense, science hasn't allowed for the possibility of its own existence.

The explanation of how knowledge can exist within nature needs a distinction between reality and information about that reality because the information can potentially be wrong although it must exist before it is proven wrong. Knowledge is true when it corresponds to reality, but it is not necessary for everything we believe in to be true. In present science everything that exists is always true. This is not necessary with information. False information can exist in nature. This requires a distinction between *knowing* and *being*. The distinction does not mean that knowing and being reside in different worlds (Plato, for instance, made the present world of being an imperfect reflection of another world of knowing). Rather, every instance of knowing has a being although that being is distinct from the knowing. Therefore, while the existence of information is true, the information itself may be false. This means that there is something physically existent although its meaning does not correspond to the facts about other things it represents.

The notion of symbols creates the possibility for interpretations about matter to exist within matter. Now, an object can hold information about another object, and while the referring object exists, the knowledge it conveys about the referred object could be false. Unlike physical properties whose existence is the truth, the existence of meaning does not guarantee its truth. I might think that the 'sky is green' and the thought exists in my brain and it can be measured physically. However, the meaning conveyed by the thought isn't true. For the brain to hold meanings there must be something in nature that can exist without being true. In classical physics, whatever exists is also true; there is no room for false things to exist, because the world only exists as physical properties. Symbols don't suffer from this problem because there are two ways to understand the symbol. The physical properties in the symbol indicate that the symbol exists. The meaning in the symbol indicates whether the claims made by the symbol are true. If a physical state is false, then that state does not exist. However, if a semantic state is false, the symbol physically exists, although its meanings are not true.

This tells us that there are two notions about truth—physical and semantic. Accordingly, there must be two notions about scientific claims, only one of which exists in classical physics. Classical physics is an incomplete description of reality because it deals with the truth about the existence of propositions, but not with the truth of the propositions themselves. If, therefore, quantum theory is indeed a final theory, then it must distinguish between these two types of truth claims. This means that quantum theory should be capable of dealing with meaning and that quantum theory would reduce to classical physics only by discarding questions about the truth of meaning. Since classical physics cannot deal with meaning, trying to reduce quantum theory to classical theory will essentially leave questions about mind and meaning outside of science.

An alternative approach to interpreting quantum theory is that it is a theory of meaning. It never reduces to classical physics, although classical physics describes matter in terms of physical properties by discarding questions about meaning in matter. The conflict between classical and quantum theories, in this alternate approach to inter-pretation, is based on the false premise that the everyday world is classical, because the macroscopic world is obviously semantic. Classical physics reduced the semantic world to a physical world, and this seemed to work because classical physics studied objects as if meaning did not matter. Since meaning arises in symbol collections, and classical physics studied individual particles, the semantic properties remained invisible. Quantum physics deals in collections rather than independent objects. Quantum collections have properties that a collection of classically independent particles cannot. This can be understood if matter is viewed as information. However, now, both macroscopic and microscopic worlds are informational; the macroscopic world cannot be classical.

If an informational conception of the macroscopic world exists, then it would free physics from the trouble of trying to reconcile classical and quantum theories. It will free us from the problematic idea that realities at atomic and macroscopic levels are basically different and need to be described by different theories. With such a view, quantum theory would not require an interpretation, because we will see quantum principles in the everyday world.

Einstein once said that science is a refinement of everyday experience. Classical concepts such as particles and fields, position and momentum, were drawn by abstracting everyday experiences into physical ideas. If today these concepts conflict with quantum principles, why would we not look back into the everyday world and find new intuitions to describe reality? The successes of classical physics and its wide application in the development of machines seemed to have convinced physicists about the mechanical nature of the macroscopic world. And this mechanical view conflicts with quantum theory because, in quantum theory there are (a) no point particles, (b) no determinism, (c) no continuous trajectories, (d) the classical ideas about the continuity of space-time are suspect, (e) the notions about the separation between matter and its state are doubtful, (f) the concepts of locality are violated, and (g) the idea that observation detects a pre-existing reality is no longer true. With such a deep conflict between classical and quantum ideologies, attempts to reconcile them at any level are not likely to succeed.

I will therefore sidestep all attempts to reconcile the classical and the quantum worlds. In my view, the classical worldview is false. It was developed to explain objects that do not describe other objects and questions about meaning were never important. The classical worldview works when we study isolated objects. By isolating objects, we minimize the effects of their semantic properties, because semantic properties only appear in collections. The classically isolated particle is an idealization, which makes us believe that the world is comprised of independent things. This idealization is incorrect when objects are collected, and the semantic properties of collections are already well-known in the macroscopic world. The bizarreness of quantum theory can be demystified if these intuitions about semantics are also extended to the microscopic world. Such an interpretation will tell us that meanings are not creations of mankind. They are rather fundamental features of nature itself.

Identifying the Intuitions

To think differently about quantum theory, we need to identify some ordinary experiences that don't fit within classical physics. These experiences can then be used as the intuitive foundation to further our discussion about quantum theory as a description of non-classical phenomena. Of course, such an approach, in general, is contentious. If X does not fit within the scope of a theory T, using X to discuss T can be flawed. After all, T may never have been intended to explain the phenomena X. However, the approach becomes necessary if X and T are very generic, or aim to be very generic. Here, T is quantum theory, which aims to describe the entire universe. Then X can be anything within the universe and the more widespread and common X is, the more relevant the need to discuss T in the light of discrepancies that arise from T trying to explain X.

A phenomenon that does not fit in the classical worldview is *knowledge*. Classical physics cannot explain knowledge for three reasons. First, every classical particle has possessed properties that are independent of other particles. A symbol, however, acquires meanings in relation to other symbols, and the meaning in a symbol does not exist outside this relation. Second, every classical particle only describes itself and we cannot interpret the particle's state into knowledge about a different particle. A symbol that represents information about another object is however at once similar and different from its reference. By representing information about something, a symbol does not become the reference although it certainly captures salient features of its reference. Third, there is another class of relations by which symbols can refer to other symbols, through what we call *naming* in philosophy. By this relation, a symbol can become knowledge about another symbol.

In this respect, knowledge is a non-classical phenomenon because classical physics is exhausted by the possessed physical properties of material objects, which are defined independently of other particles and there cannot be anything in an object that is not entirely about that object. Can we now postulate that since quantum theory is a non-classical theory, and since knowledge is a non-classical phenomenon, quantum theory may be *potentially* a good explanation of knowledge? I understand that the recognition that knowledge and quantum theory are both non-classical doesn't imply that they are related. And I'm not implying that the non-classical nature of knowledge should be reason enough to convince us of its relation to quantum theory. I'm only suggesting that this could be a working hypothesis, until we have worked out its ramifications.

It is also important to note that I am not talking about knowing the meaning of quantum theory, which is a philosophical question. I am instead talking about using quantum theory to explain the phenomenon of meaning, which is a scientific question. A theory of reality must not just explain reality, but also knowledge about that reality because knowing reality is part of reality. Current science assumes that knowing is different from matter. In Greek times, this was the divide between the Platonic world of forms and the everyday world of things. After Descartes, ideas have a place in the world, but outside matter—into another substance called 'mind'. From Plato to Descartes, we brought knowledge from the other world into the present world. We now need to move another step forward in the same direction as Descartes and bring knowledge inside matter.

Modern advances in biology and neuroscience have shown that

much of what we earlier regarded to be features of the mind are in fact properties of the brains, which are physical objects. These advances have created the need to understand knowledge in terms of matter. Ultimately, this is a problem of science coming to terms with the existence of knowledge. If quantum theory is a fundamental theory that governs everything, then it must govern our brains as well. Knowledge in the mind is at least correlated to some physical states in the brain. In order for quantum theory to explain the brain it must be able to explain the phenomena of knowledge in the brain. In fact, if the brain is studied symbolically, then its existence—as physical particles—is not *empirically* different from the meaning represented in it because the meaning is in the relation between the symbols. By observing the symbols and disregarding their relations we can study the brain physically. But by viewing the same brain as relations between symbols, we can treat it semantically.

Any theory of knowledge needs a reconciliation between reality and its meaning. Quantum theory as a theory of knowledge must also therefore require a reconciliation between objects and their meanings. And this is a non-classical requirement for quantum theory, assuming that the theory is an ultimate theory about reality.

If quantum theory governs the brain, then reality and its knowledge are governed by the same theory. This is similar to the relation between a picture and the landscape it displays. We might say that a picture is an artist's impression of reality, although it is not reality. But the picture is also part of reality, when you expand the definition of reality. You can also have someone recursively draw a picture of the picture! The brain similarly can picture reality and be part of that reality at the same time: both the knower and the known are real. Accordingly, they should be explained by the same theory. The key non-classical question is: What are the requirements for a theory that describes pictures of objects in addition to objects? Note that pictures too are objects, although not in the same sense.

If quantum theory does not explain knowledge, then there must be another theory that explains it. This quickly leads to an infinite regress. Now we need theories that will explain the brain's capacity to know quantum theory, and then theories that explain the brain's capacity to know that theory about the brain, and so on, ad infinitum.

We might then conclude that we cannot know everything.

If, however, quantum theory is a theory of knowledge, then infinite regress is avoided and everything is knowable. However, in that case, there has to be something within quantum theory that explains knowledge. Interpretations of quantum theory are part of what we know. By such knowing, we form our view of quantum theory. If quantum theory can explain knowledge, then it can also explain the manner in which we form interpretations of the theory, which are various representations of its formalism. Interpretations are thus not irrelevant to quantum theory. As philosophical problems, it is easy to deride their significance by assuming a fundamental divide between the mind and the world. As scientific problems, interpretations place new demands on any fundamental theory of nature. My approach to interpreting quantum theory is to use knowledge as a non-classical phenomenon to be explained by quantum theory. Now, we are not just trying to explain reality but also knowledge about that reality. If quantum theory can explain knowledge, then quanta are reality used to represent reality.

This approach to formulating an interpretation of quantum theory differs with the traditional philosophical approaches. It makes the philosophical problem of interpreting reality into a scientific problem of explaining the mind. The requirement for interpretation can then be restated as follows: A theory of reality must explain *pictures* and not just *objects*. Currently, all interpretations of quantum theory think of it as a theory about objects, as an extension of classical mechanics that deals with objects. My focus here is to provide an interpretation of quantum theory as a theory of pictures.

What is the key difference between a picture and an object? An ordinary object is a collection of facts—mass, charge, length, momentum, etc. A picture has these properties as well, although the picture also *represents* another object's facts through its facts. When you look upon a picture as an object (merely analyze its mass, charge, shape, size, momentum, etc.), you don't get its meaning—the things that it refers to—although the meaning is part of the picture. To know the meaning you need another *mode* of knowing the object, where we attempt to understand not what an object *is* but what it *represents.* All of modern science is the study of what a thing is and not what a thing represents. The ability to represent has to exist in matter, for it to be able to explain the mind. However, representations are not necessarily associated with the mind. Ordinary objects can also be representations; pictures hanging on the wall are examples. The task of interpreting the theory therefore need not be associated with the mind. It can also be associated with a class of things that represent other things. Minds are amongst this class, although minds don't exhaust the class of things that refer to other things.

The Consequences of Symbolism

There are three clear consequences of treating reality as symbols instead of physical particles. The first is that we get a new way of sub-suming classical physics within quantum physics. A symbol gives us information, which can be about other objects, or it can be about itself. A classical ensemble informs us only about itself. In that sense, it is a symbol for itself, although we don't treat that ensemble conceptual-ly—e.g. a 'chamber of gas'. We rather treat it as a collection of point particles which are devoid of everyday concepts like 'chamber' and 'gas'. Symbols can additionally inform us about other things, in which case they would not fit within classical physics, but they could fit into a new physics that directly deals with information instead of classical particles. The second consequence of viewing reality as symbols fol-lows from the first. If a symbol can represent reality, then it must also be able to represent non-classical properties, besides classical proper-ties. For instance, a book describes the world in terms of non-classical properties like color, shape and size, not in terms of classical prop-erties like mass, charge, momentum, etc. If nature is symbolic, then it must be able to represent both classical and non-classical proper-ties. This is consistent with the fact that our brains represent reality as non-classical properties and not as properties that physics attri-butes to reality. The third consequence is that if symbols can represent non-classical properties, and symbols can describe themselves, then a symbol that describes itself non-classically will have non-classical properties. This is more profound than might initially seem. It implies that matter has the ability to convey information about its own color, taste and smell, which are non-classical properties. Note that I am not saying that matter *is* color, taste and smell. I am only saying that matter can convey information about how we may perceive it.

Early in the history of empiricism, philosophers drew a distinction between primary and secondary properties. Secondary properties are how we perceive the world, and they include sensations such as taste, smell, color, etc. Philosophers argued that the way we perceive the world is a human way of looking at reality, which is generated by our minds. Nature itself, however, exists as primary properties, such as length, mass, charge, momentum, etc. which are properties that we can measure in relation to other objects, not in relation to human observers. In this view, nature exists as primary properties, but it is *converted* into secondary properties by our observational apparatus. This view creates the following important problem in science. We can say that an object's physical state is real, but we cannot say that its redness or sweetness is real, because redness and sweetness are facts about our perception, not about reality. The claim that the apple is red is false. It is truer to say that we *perceive* the world as red and apple. This implies that everything we see is an illusion created by the mind. How the mind creates such illusions requires a new type of explanation.

This divide between reality and its perception creates serious problems for a physical theory of the mind. First, how do real physical states in the brain lead to illusions about color, taste and smell? Second, what makes the brain think that this illusion is reality and that the world is red and sweet instead of position and momentum? Third, if the physicist is sure that the brain is always an illusion about the nature of reality, then how did he come up with reality in the first place—isn't his brain supposed to be under illusion as well?

These problems have not yet been solved, and I don't believe they will ever be solved within the framework of current physical concepts. They can however be solved in the symbolic view if we say that symbols in reality represent color, taste, smell, etc. Note that we don't have to say that symbols *are* color, taste and smell. It is sufficient to say that they convey *information* about color, taste and smell, using which the observer creates sensations. When we say that the apple is red we are using the same *words* to describe reality and our experience, but the words have different meanings. In the mind, redness represents a sensation. In the external world, redness represents a symbol of redness. The symbol is objective, and the sensation is subjective. They can be described by the same words, but they are different. The simple advantage of this scheme is that our theories about nature can be validated through perception, because the words that describe perception can also describe reality. In current science, physical concepts need to be *mapped* to perception, when this mapping is under question and yet to be explained. For instance, how do we know that the emission of a photon will create the perception of color in an observer, when we haven't yet explained the process by which eyes perceive the photon as color?

Imagine that you are standing at a money transfer teller and you wish to transfer Euros from a bank in one country to a bank in another one. Assume also that the two banks, which transact money between each other, always transact in US Dollars. The originating bank thus converts Euros to Dollars which are then converted back into Euros at the destination bank. If you have never seen the Dollar, this scheme will be totally transparent to you. The conversion scheme could also involve any currency whatsoever, not just Dollars. For instance, the originator may convert Euros to Rupees, transfer them as Rupees, and then convert them back to Euros at the destination. What you perceive is Euros, but the reality beneath is Dollars or Rupees. A similar problem plagues scientific concepts. The experimenter prepares a state and then observes that state. The state preparation and observation involve human concepts. In between, the physicist converts human concepts into physical concepts and then back to human concepts. There is a certain amount of arbitrariness in these conversions, which goes unnoticed as long as experiments are successful, similar to how you ignore the currency in which money is transferred as long as it is transferred. The physicist's view is that if we can successfully transfer Euros from one place to another, then Dollars are real. Would it not be much better to say that if you can transfer Euros then Euros must be real?

The Primary-Secondary Property Divide

The current divide between primary and secondary properties is not just a difference between experience and reality, but also a difference in the *language* we use to describe experience and reality. Reality is described using primary properties such as mass, charge and momentum while experiences are described using secondary properties such as color, taste and smell. The divide between experience and reality does not necessitate a linguistic divide, although early empiricists instituted a divide. They claimed that if experience and reality are different, then they must be described by different *words.* By using different words to describe experience and reality we create additional problems in describing the brain.

For instance, if the external world exists as frequency, but it is perceived as color, then the brain *is* frequency but it *represents* color. How does the brain know how to convert frequency into color? If color is a human way of seeing matter, and the brain is material, then, to represent color, this human way must enter matter. And if the brain can in fact represent the property of color, then why can't we suppose that the external world itself represents color (since the brain and the world are both comprised of similar physical properties)? Note how the problem of explaining the perception of the observer is not a problem of the mind in a physical theory of mind. It becomes a problem about the nature of the physical state itself.

The problem of knowledge and meaning is a problem of representation in matter whereby matter can convey something other than what it is. In the context of a particle representing the state of another particle, we only require primary properties to be represented. But, in the context of the brain, we need secondary properties to be represented as well. The transition from primary to secondary properties now needs to happen within matter itself.

This creates the problem of how some primary property can be a representation of a secondary property. And if such a representation can indeed exist, then we are wrong in supposing that the material world is only primary properties. The only way out of this conundrum is to dissolve the distinction between primary and secondary properties *at the level of language.* We should now be able to say that the primary properties themselves *represent* secondary properties. The red apple in front of me is not mass or momentum that I perceive as redness. Rather, the apple itself is a representation of the idea of redness. During perception, this information about redness is represented in the brain as a symbol, and subsequently perceived as the color sensation of red. Both the external property of redness and its representation in the brain are objective and can be known by others through observation. The sensation of redness is subjective and can't be sensed by others. However, the same *word*—redness—is used to describe reality and its perception.

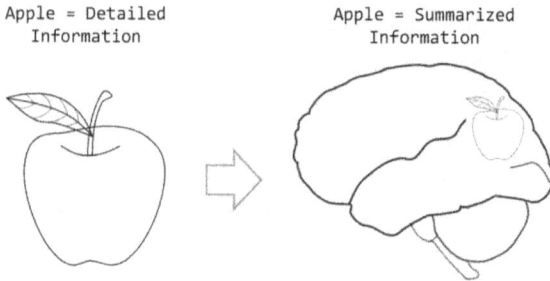

Figure-1 Two Representations of an Apple

The general thinking on perception is that the external world is physical while the mental representation is ideas. When the physical world and the mental world are described by two different *languages*—i.e. primary and secondary properties—then we create the explanatory gaps of how the physical world becomes ideas. This problem can be solved by dissolving the linguistic divide between the external world and the mind. For instance, the external world is information about a red apple which is very detailed, and exists as atomic objects. During perception, information in ensembles of such atoms is compressed into a summarized representation in the brain. This would mean that the real apple is a thing, but that thing encodes concepts. During perception, we don't transfer the thing, but we transfer the concepts. The real apple too must be described therefore in terms of concepts; in this case, as symbols of red, round and sweet. The external world is now a *symbol* of red, round and sweet and the brain too is a symbol of red, round and sweet. Both these symbols are objective and can be studied as physical properties. However, we are better off describing these symbols in terms of meanings.

Take the difference between tones and notes, for example. A tone is a frequency, amplitude and wavelength. But a collection of such tones also become notes. A quantum eigenfunction[1] is similarly a vibration and can be studied as frequency, amplitude and wavelength. But in relation to other such vibrations, it could be a note. Now, one might argue that if the tone becomes a note, then what's the benefit of using notes instead of tones? The difference is that the tones *underdetermine* the note; the same tone may represent a different note within a different collection. Therefore knowing the tone (frequency, amplitude and wavelength) does not fix the note. Rather, the tones must collectively define the notes. If causality is based on notes instead of tones, then knowing the tone will underdetermine the note and hence it's true causal effect. If causality is based on notes, then we must describe the sound vibrations as notes rather as tones. This would collapse the divide between primary and secondary properties. The tone is *objectively* the note through its contextual relation to other tones. This objectivity cannot be reduced to a *single* object but to a collection as a whole.

Symbolism in matter collapses the primary-secondary property divide, as far as the linguistic description is concerned. This in turn implies that the language of science—where we described nature only in terms of primary properties—must change as well. Instead, the language of science must use the same *words* as the words we use to describe our sensations. But these words will not represent sensations. They will instead represent physical states. In short, physical states now need to be described using the same words as we currently use to describe sensations. We now need to say that matter *represents* color, taste, smell, etc. instead of saying that it *is* mass, charge, energy and momentum. The notions of color, taste and smell need to be objectified in matter in the sense that the physical state itself can represent color, taste and smell. But the language of natural description would be drawn from the observer's experience of the world rather than from considerations of what primary properties may exist 'behind' the perception of secondary properties.

Accordingly, the notion of state measurement must change from seeing the world in terms of mass, charge, momentum, to seeing it in terms of color, taste and smell. This description of reality is similar to how we derive meanings from reading a book. The symbols in a book are physical states, but these states are also representations of meaning. What exists is the physical state, but what we derive is meanings. In a similar way, if nature is symbolic, then physical states and meanings will be two ways of describing the same reality.

In the physical description, there is a difference between the language used for reality and experience. In the semantic description, reality and experience employ the same language. Thus, we might say that we see red because the external world represents red. At least sometimes, then, we can say that when I see red, the world is indeed representing red. Symbols may also convey information about other symbols, and by experiencing these symbols we will be acquainted with a reality different from the one that is experienced.

Symbolism introduces a distinction between the external reality and the perceived meaning, although both are described by the same words. Current physics, through its distinction between primary and secondary properties also distinguishes between reality and meaning, but uses different words to describe them. Reality is length, mass, charge, momentum, etc., but meanings are red, round and sweet. This creates the problems of mind and meaning, because the distinction between reality and meaning has to be bridged somewhere in the brain, by bridging two different languages. The symbolic view of nature solves this problem by using the same words to describe both reality and meaning. This does not equate reality with the *experience* of meaning, but only as an *objective symbol* of meaning. Symbolism only implies that reality is described using the same words as the meanings about it. Since meanings are red, round and sweet, reality too must be red, round and sweet, instead of current primary properties. Now, the only correct way to think about reality is that it is *information* about our sensations. If I see a red apple, then the apple is indeed representing red (and not mass or charge) but it is not red as my sensations but as semantic objective information that encodes the idea of redness within matter.

Understanding Quantum Problems

The properties in a classical physical object are independent of the properties in the other objects. Quantum properties are different because even the physical properties in a quantum object are defined in relation to the physical properties in other quantum objects. This fact suggests that we should treat the quantum physical properties like notes, although we would have treated the classical physical properties as tones. It also implies that even though quantum theory borrows the vocabulary of classical physics (i.e. using words such as position, momentum, time, energy, angle, angular momentum), the meanings of these words are different in the two theories. The same words in quantum theory can denote meanings while they would have represented physical properties in classical physics.

If quantum objects denote meanings through their physical properties, then we can provide a new interpretation to the *succession* of these objects. In the way that a succession of notes represents a musical composition—and hence a more complex meaning—the succession of quantum events represents a complex meaning. It is possible to now conceive of new laws that describe how simple meanings combine to produce complex meanings. These laws will predict the succession of quantum events, which current quantum theory cannot. In current quantum physics, the succession of quantum events is described probabilistically similar to how a musically ignorant person can measure a musical composition and only obtain the probabilities of various tones. For the musically ignorant, the succession of tones does not represent a complex semantic object. For the musically savvy, who interpret the tones as notes, the succession of notes is a more complex meaning.

By shifting our viewpoint, such that we interpret the properties of a quantum object as semantic properties, we open quantum theory to new laws of semantic combining. These new laws will complete quantum theory because we will *read* the succession of quantum events as the succession of meaningful symbols. The semantic viewpoint is therefore not merely a philosophical reinterpretation of the current mathematical formalism, but a vista to formulating new types of laws that can overcome the current incompleteness.

We should note that the relation between meaning and physical properties is contextual and a physical property does not universally translate into a meaning. Each context (quantum ensemble) defines a unique mapping between meaning and physical property. This means that the physical property *underdetermines* the meanings and so long as we describe the world as physical properties, our descriptions will be incomplete, if the world is in fact semantic. Thus, position=X is incomplete although color=red is complete. Since current quantum theory describes reality in terms of physical properties, it is incomplete. It can be completed if we can reinterpret the physical properties as symbols of meanings because that reinterpretation will open the doors to new types of laws.

The same physical property across two different ensembles will, in general, represent different meanings. This is similar to saying that the same physical token or squiggle will denote different symbols in different contexts. The physical property is therefore not meaning. Rather, the meaning can be *expressed* by some physical property. That also implies that the mind is not equal to matter, although the mind can be expressed through matter. The same idea can be represented through different physical states in different contexts. In a given context, we can derive the idea from some given token, but the idea does not reduce to the token, since there are always other contexts in which the same idea can be expressed differently.

This approach to interpreting quantum theory highlights the nature of incompleteness in the theory. The incompleteness is not in the observations themselves, but in the explanation of the observations. Depending on the kind of reality we postulate underlying the quantum observations (semantic vs. physical), the laws are quite different. The physical approach clearly cannot predict or explain the succession of quantum events, which makes the theory probabilistic. The semantic approach can explain this succession, although it requires new laws. It is a common misconception today that quantum theory is complete because of Bell's Theorem[2], which shows that any theory that adds new variables or observables to the theory will contradict its predictions. To understand why this is a misconception, the difference between tones and notes is illustrative.

For instance, to interpret a note from a tone, no additional hidden variables or observables are required. The meaning—the note in this case—is neither hidden nor an additional observable. Rather, we derive the note from the same observable—the tone. However, the note is not identical to the tone because in a different context the same note can be expressed by a different tone. The note is a contextual interpretation of the tone, and it depends on the collection of tones as a whole. There are two ways to understand the tone—as physical properties and as a symbol of meaning. If we look at the tone as physical properties then we cannot predict the succession of tones. But if we look at the tone as a note then we can formulate laws to predict the succession of notes as a composition.

Figure-2 Symbols Convey Meanings and Percepts

The knowledge in a symbol does not reduce to the symbol's physical properties, although the knowledge does not require new observables. Consider the picture in Figure-2. It has a certain shape, size, color, all of which can be described in terms of physical properties. However, this picture also denotes the meaning "fragile," which is not described by the physical properties. As physical properties, the picture is a thing-in-itself, but as meaning it is a thing about things which will be called fragile. Fragility itself can be described in terms of other physical properties such as the strength of materials, malleability, ductility, etc. However, the picture itself has little to do with the strength, ductility, malleability, etc. The picture itself may or may not be fragile to denote fragility.

This difference between what a thing *is* and what it *represents* holds the key to the solution to the quantum problem. Symbols can be seen in two ways: physical (what a thing *is*) and semantic (what it *represents*). Current science attributes causes to what a thing is and not to what it represents. This is because we measure the world instead of reading it. To overcome incompleteness, physics needs to shift from classical to semantic observations. Obviously, now, we must also attribute causality to the meaning in the symbol rather than to its physical properties. This has important ramifications about the nature of physical law

as conceived in science. Unlike current laws, which are based on physical properties, semantic laws will be based on the meanings. These
laws will prescribe rules for combining atomic memes to construct
complex memes.

Quantum Theory and Perception

Classical physics entailed a theory of perception in which the world
exists as primary properties (length, mass, charge, momentum, etc.),
which the observer perceives as secondary properties (color, taste,
smell, etc.). The conversion of primary properties to secondary properties, is in this view, a problem for the biologist or neuroscientist,
not the physicist. The neuroscientist addresses this problem by saying
that the brain obtains a representation of the external world, which is
somehow perceived as the sensation of color. This explanation of color
perception, however, involves two fundamental unsolved problems.
First, there is nothing in physics that tells us how some brain state
can be a *representation* of the state in the external world, because all
states pertain to the measured object, not to another object supposedly represented in that object. Second, since the external reality is
primary properties, even if the brain creates a representation of reality, it would be a representation of the primary and not the secondary
properties. For instance, when we perceive color we believe the external world is the frequency of light. If the brain creates a representation
of the world, it would be a representation of the frequency of light
rather than its color.

Classical physics implies a theory of perception in which matter
cannot represent the state of another object besides itself and even
if it were to represent it, the represented state would pertain to primary and not to secondary properties. How primary properties in the
brain represent other primary properties in the external world, and
how primary property representations in the brain become secondary property experiences are a mystery in this theory.

Developments in quantum theory and the biology of perception
have revealed additional facts about brain chemistry, but have not
substantially altered the classical view of perception. The current

view states that photons from the external world enter the eyes and are absorbed into cones. There are three types of cones, which absorb light in the frequency ranges that we perceive as red, green and blue, and cones thereby create an RGB representation of the light's color. The cones then release electrons, which travel to the brain through nerves, thereby carrying the RGB representation of the color created in the eyes. This description of the process of perception offers no new insights on the two fundamental questions about perception that classical physics left unanswered. First, how do electrons traveling to the brain *represent* RGB color distinctions, when fundamentally all electrons are identical particles? Second, how does the brain see the electron as a *representation* of reality when all that can be obtained from the electron is its position and momentum?

To see some objects as pictures of other objects, science needs a new way of interpreting nature, in which the properties of pictures indicate properties of intended objects. This view of nature is not provided by classical physics, or by the current quantum theory.

The symbolic view of quantum theory can help demystify these problems. The electron's eigenfunction can be viewed as a symbol of red, green and blue in addition to the *name* to which these properties pertain (I will later describe how different quantum properties can be interpreted to encode names and meanings). In effect, if the brain symbol represents the proposition 'the sky is blue,' then the word 'sky' is a name that refers to some external object, and the word 'blue' represents the meaning of the above named object. This addresses the two problems of representation described above. The quantum in the brain is a symbol of blue and not a symbol of frequency, and the quantum refers to the intended reality by a name.

Of course, the quantum in the brain is not different from the quantum in the external world. Therefore, if the brain quantum encodes names and qualities, then the external quantum must also encode them. The difference only needs to be in the intended names: the external quantum should name itself (assuming for the moment that it is not itself a description of another reality), while the brain quantum names the external quantum. As far as qualities are concerned, both brain and external quantum must represent the same quality—e.g., the quality of blueness. This means that the external quantum is a symbol of the property of blueness, which we currently measure as momentum, angular momentum and spin. The photon that carries information about the external quantum must encode information about blueness; the cone that absorbs blue photons selects the photon based on its information content (i.e. blueness) and the brain electron obtains information on blueness.

The solution to the problem of perception thus entails a radical revision to our notion of reality. Specifically, the external world is not primary properties, which is then perceived as secondary properties in the brain. Rather, the external world encodes blueness, which is transported by light as blueness, which is selected by eyes to create a signal of blueness, which is represented as the sight of blueness in the brain, and then ultimately perceived as the blue sensation.

The light from the external world carries information and intentionality about the world into the eyes. The eyes transfer information and intentionality about the world and the eyes into the brain. The brain thus believes that it sees the world as blue through the eyes. Note that the brain doesn't just think that it directly sees the world. The brain believes that it sees the world *through* the eyes. This is more pronounced in the case of touch, taste and smell, where we are acutely aware of the role of the senses in obtaining sensation. The sensation of heat on the skin is not just attributed to the external hot object but also to the skin perceiving the heat, which is also experienced at a specific point on the skin. Similarly, the sensations of taste and smell are attributed both to objects and to the senses—in this case tongue and nose—involved in the perception.

Neuroscientists today construct the following classical picture of the process of perception. In this picture, the body and the brain are connected through neurological wires and the brain knows which wire leads to the skin and which to the eyes. In a sense, the brain has a 'map' of the body, by which it attributes sensations to different body parts. This view has an empirical basis in the observation that different parts of the brain perceive sight versus smell. The neuroscientist theorizes that this happens because neurological wires from the nose and the eyes terminate on different parts of the brain. To know that my leg is itching and not my hand, the brain only needs to refer to the different wires through which the signal is coming. But this view of perception misses the fundamental point: How does the brain know which wire represents which *type* of signal and which wires are coming from *where* in the body? The signal from the eyes and the nose are not the same type of signal, even though they are carried through the same types of wires and ultimately produced by the transport of physically identical electrons. What makes one wire the signal of smell and not of sight if, fundamentally, these wires and the electrons in them are all similar? Similarly, if all wires and electrons are fundamentally similar, how can the brain label the origins of these wires to different body parts?

This problem cannot be solved if all we attribute to the quantum is classical physical properties, because of the same two reasons identified earlier. The electron has to represent a *type* and it has to *refer* to its source. The source is identified by a name (hands vs. feet) and the signal is identified by a type (hot vs. cold). The brain needs to know the type of sensation (e.g., hot vs. cold) and the location from where the sensation is being created (hands vs. feet). These two requirements create the problems of naming and meaning within quantum theory. If the brain is fundamentally only receiving the position and momentum of electrons, it has no way of knowing that some electron represents heat instead of cold. It also has no way of knowing that the heat signal is from a leg instead of a hand[1].

The semantic view, however, can demystify the problem of perception in which quantum objects denote both names and meanings. The stream of electrons is not now a stream of physical particles. It is rather a stream of symbols. Each symbol in the stream has naming and meaning properties; in effect, each symbol is a proposition about some fact in the external world. For these facts to be derived from the external world, the external world itself must be capable of naming and meaning. Now the fact that we see red color is not a creation of our senses or minds. It is rather an objective fact about the world; the world itself encodes the meaning of redness.

Semantic Concepts of Change

The semantic view substantially alters our ideas about change. If matter is physical particles, then change is motion. If, however, matter is

symbols of meaning, then change represents an evolution of knowledge! While the motion of classical particles was continuous, the evolution of knowledge must involve changes through leaps.

Quantum theory currently provides no succinct notion about
change. While classical physics equated change with motion, in quantum theory, there is no motion. This is because motion requires the
continuity of space-time, and a continuous succession of space-time
object states, both of which are impossible in quantum theory. First,
space and time cannot be divided any smaller than Planck's length;
if we cannot conceive infinitesimal lengths and durations, then we
cannot speak of motion in the classical sense. Second, and even more
importantly, the position states of quantum objects are themselves
discrete, and two quantum states cannot be infinitely close; a quantum object can only jump from one state to another. When a quantum
particle jumps between states, we cannot say that the *same* particle
has moved to a new state, which is so crucial to the idea of motion in
classical physics (a moving object only changes its position without
changing its identity). These two problems create serious challenges
for a quantum conception of change.

Quantum theory requires a revision to the idea of change. Since this
change takes place in space and time, in which classical physics modeled change as motion, revisions to the idea of change entail a revision
to the idea of space and time itself. The notion of space-time must be
altered, from a container of objects to a container of symbols. With a
new view of space-time, the idea of change will also be altered from
the motion of particles to the evolution of knowledge.

If reality is symbolic, then the space in which it exists is semantic. That is, different locations in space represent different meanings.
In a semantic space, all changes must be discrete because knowledge
always evolves discretely. If the space-time in which matter exists and
evolves is semantic, then change must be described as the evolution
of knowledge. Particles in classical physics are physically distinct and
this distinctness is known by their unique locations, although they are
of the same *type*—i.e. they are all particles. In a semantic space, distinct locations in space don't just indicate a physically distinct object
but also a conceptually distinct *type* of object. Thus, there are many
identical particles in classical space, but no particle or symbol can

be identical in semantic space because all locations identify different types of meanings.

Classical physics has four primary symmetry properties—the homogeneity and isotropicity of space and homogeneity and isotropicity of time. I will later argue that these symmetry properties can be given a semantic interpretation, encoding four different aspects of symbols through quantum objects, and these aspects can represent *descriptive* information. Corresponding to these four symmetry aspects of space and time are four conserved properties—namely, momentum, energy, angular momentum and spin—which too can be given a semantic interpretation, and these encode *programmatic* information. The descriptive and programmatic aspects of symbols are complementary and this is represented in quantum theory by the complementarity between space-time properties and conserved properties. A quantum object can thus be viewed as a symbol with both descriptive and programmatic information. The only radical revision to current physics needed to incorporate a semantic view of quantum theory is to treat space and time as semantic rather than as physical containers. In current physics, all dynamical properties of objects are treated as quantities. In the semantic view, the same properties will now represent types.

This viewpoint helps us answer fundamental questions about the origin of space-time, and hence of the universe. The answer is that locations and directions in space and time represent distinctions in meanings and distinction locations in space-time (events) can therefore be produced from meanings, if the meanings exist prior to space-time. The origin of these meanings will remain a metaphysical question from the standpoint of space-time physics, although the meanings themselves can be studied in space-time.

Overview of the Semantic Interpretation

SI claims that current quantum theory is not an ultimate theory, because it makes probabilistic predictions without giving a causal picture of change. The classical view of change as motion, causality as deterministic force laws, and space-time as a continuous container is

false in quantum theory. This creates interpretive difficulties in understanding quantum theory vis-à-vis classical physics. The difficulty can
however be resolved by supposing that the everyday world is not a
classical world; it is rather a world of information. Classical physics is
a physicalist idealization of a semantic world. In this idealization, the
discreteness and contextuality of information is ignored to create the
impression of a continuous space-time containing infinitesimal particles, each of which is independent of the other particles in space-time.
Quantum theory is fundamentally incompatible with this idealization.
The quantum problem can be solved by returning to the everyday
world and restoring those informational primitives that were eliminated from classical physics.

The widespread idea that quantum theory is only about the atomic
world is therefore false. Quantum theory is a theory of both the
atomic and macroscopic worlds. The conflict with classical physics is
however real because this theory rejects meanings. We need a more
forceful rejection of classical concepts than has so far been done. The
quantum-classical conflict (which appears as the famous measurement problem) can never be solved. However, it is possible to give a
quantum-compatible understanding of the macroscopic world, making quantum theory a universal theory.

Current quantum theory does not provide an adequate view of
quantum objects or a picture of how change occurs. It continues to use
classical object-concepts (particle and wave) and a classical notion
about change (motion) even though all fundamental principles on
which the classical view was founded are false in quantum theory. To
bridge the gap between classical and quantum theories, many other
notions such as uncertainty, complementarity, non-locality and collapse have been added and all interpretations try to interpret these
principles. In SI, these principles are explained as shortcomings of the
attempt to maintain classical views about matter. Now, uncertainty is
a consequence of a symbol's extension, which enables the symbol to
represent type distinctions. Non-locality is a logical relation between
symbols in an ensemble. Wave-particle duality is a consequence of the
fact that a quantum represents meaning through a *vibration* (like the
sound symbols in speech), but that vibration is also a particle—i.e. a
unitary symbol. In SI, the classical picture of reality is rejected and

not enhanced. The goal of SI is to show that a semantic view of reality explains the unintuitive ideas in current quantum theory in a way that we can substitute the current theory with a new one that is free of its riddles.

Classical physics is centered on the *motion* of objects, and this defines our ideology of change: all change is motion. Quantum theory requires us to think of change without thinking of motion. We have to conceive of a new picture of change that does not involve motion. Quantum non-locality involves action at a distance, without motion. Quantum measurements involve detections that transfer mass and energy although there is no trajectory. Uncertainty implies that we cannot draw a path. The crux of the quantum problem vis-à-vis classical physics is that we are trying to imagine *change* in terms of *motion*. What if there are changes that don't involve motion? How would we visualize changes in matter without motion?

Like falling apples was an important metaphor for change in Newtonian physics, we need a metaphor for change in quantum theory. The following everyday example I think will prove helpful.

Imagine you are looking at a LED sign built out of small LED lights. Assume also that the sign uses a controller that turns the lights on and off in a sequential and coordinated fashion, giving the impression that the lights are moving although nothing is actually moving. The key difference between the "motion" of the LED light and the motion of a classical particle is that a moving particle is the same particle but the moving light is in fact a different light. The identity of a classical object in motion remains unchanged; the *same* particle moves to a new location. The identity of an LED light in motion changes continuously; the LED at a new location is a *different* LED. In the case of the moving LED light, there is change but no motion.

The LED analogy is useful for understanding the quantum problem. Like the LED light's apparent "motion," quantum theory also approximates classical particle-like motion, although there is no moving particle. The quantum object at position X is a *different object* than the quantum object at position Y and we must not think that the same object *moves* from X to Y (preserving its identity). We must instead suppose that an object at X disappears and the object at Y appears. The appearance and disappearance are not required to *individually* preserve matter and energy, if multiple such transforms can *collectively* preserve matter and energy. This compels us to view quantum changes collectively rather than individually. Such changes can be discrete and, to go from X to Y, a quantum object doesn't need to pass through all the intermediate states. Furthermore, we don't have to think of conservation of energy, mass, momentum, spin, etc. at the level of individual particles. These particles can undergo discrete transforms, but the system is collectively conserved.

We can revise our notion of change to object transformation, which isn't a new thing for mathematics. In fact, even classical theories of motion employ discrete transformations such as $x_{n+1} = x_n + \delta x$. There is however an incorrect assumption that Newton made in his calculus: he assumed that δx can tend to 0. This assumption allowed the idea that the *same* object moves through different locations, and the mathematical idea of transformation was converted to the mechanical idea of motion. In quantum theory, δx is no longer 0 and this contradicts the basis of the assumption regarding the motion of objects. However, it still does not contradict the transformation equation $x_{n+1} = x_n + \delta x$. We have to say that the object at x_n is transformed into the object at x_{n+1}. Object transformations require the ability to associate different object types with different locations in space. A location in space is therefore not just a property of an object; it also defines the object's identity. Objects in different locations are different types of objects (in addition to being different objects). The additional novelty is that a *collection* of transforms must occur at once such that we cannot describe the motion of a particle in isolation. We must rather describe change in which matter and energy are being *redistributed.* Thus, for example, a particle can lose mass and energy if another particle simultaneously gains mass and energy. How does one particle know how to exchange energy or mass with another particle? This turns out to be a fictitious problem arising from the legacy of classical physics, where we treat an ensemble as a *collection* of particles. It becomes meaningless if the ensemble is a *macroscopic* object. The change is occurring at the level of the macroscopic object, which causes redistribution of energy at the microscopic level. Hence, you cannot describe the microscopic changes without reference to the macroscopic change; the macroscopic change in fact precedes the microscopic changes. In short, we need to reconcile the macro and micro domains by giving priority to the macroscopic changes.

An intuitive example might be helpful here. Think of a biological ecosystem comprising flora, fauna, and geographical environment. In the reductionist view, we think that the smallest particles undergo change, which aggregates into macroscopic changes. In the new way of thinking, we would say that the ecosystem undergoes change, which then percolates into the microscopic changes. The dynamics of the entire system must be primarily described in terms of the eco-system change which causes microscopic realignments, or what we can call the redistribution of matter and energy in a system. This is the answer to the paradoxes associated with 'quantum holism'; the answer is that the evolution of the whole causes the evolution of the parts. If we disregard the evolution of the whole, we fail to grasp how discrete changes occur at the microscopic level. But if we describe the macro evolution, then the micro changes are also understood.

To distinguish different locations in space, we need some proper-ties by which to distinguish one location from another. In SI, a loca-tion represents a *meaning* and changes to the object's location also change the object's meaning. Motion in space is therefore the evolu-tion of knowledge. This evolution requires new kinds of laws than the kinds of laws in current science. Quantum theory can be completed not by adding hidden variables or new observables but by changing the nature of space, which is neither hidden nor material. A new the-ory must be created to deal with the new notions about space. This theory will describe how change to an object's location is change to its information content or meaning. An object becomes a distinct concep-tual *type* of object when it moves into a new location.

Planck's constant $\hbar$ represents the smallest possible information *distinction*. When $\hbar$ becomes 0, the size of distinction is 0—which indi-cates lack of information or informational diversity—and we end up with classical physics where all particles are of the same type. In the case of knowledge, however, $\hbar$ is not 0, because space has informa-tional properties and objects at different locations are different types of objects. This fact applies even to macroscopic objects and not just to atomic objects as in the current theory. Planck's constant is therefore not 0 at the macroscopic level. Rather, the informational properties from the everyday world can be used to formulate fundamental theories about atomic information.

To summarize this brief introduction to SI, an everyday example would prove useful. Think of how classical physics would describe a person walking on the street: a person is a particle whose identity is unchanged throughout the walk. Per quantum theory, the bodies at the start and the end of the walk are *different bodies*! Within SI, the old body disappears, and a new body appears. You cannot thus say that I have the same body from birth, to childhood, to old age. The body is changing every moment, and you are getting a new body. This change is not merely change to the *properties* of classical particle that retains its *identity*. It is rather a change both to the identity and the properties. As a macroscopic quantum object, the body is a fundamentally different object. The different stages in the "motion" of the body are like blinking LED lights whose continuity is only an illusion that comes about when we ignore Planck's constant, $\hbar$.

2

The Quantum Problem

Quantum mechanics is certainly imposing. But an inner voice tells me that it is not yet the real thing. The theory says a lot, but does not really bring us any closer to the secret of the "old one." I, at any rate, am convinced that He does not throw dice.
—*Albert Einstein*

Introduction

Quantum theory poses many interpretational problems. For the sake of discussion, I have divided these problems into six distinct areas: (a) Discreteness, (b) Uncertainty, (c) Probability, (d) Non-Locality, (e) Wave-Particle Duality and (f) Irreversibility. These are all distinct problems in themselves, but behind the multiplicity of quantum problems, there is one key issue that lies at the root of all. The root of all these problems is that quantum theory lacks a clear conception about the nature of the reality it describes. The term *quantum* apparently refers to an *object type* although the nature of this object type isn't well understood. Current quantum theory still thinks of reality in terms of object concepts that classical physics created (particles and waves), although these concepts cannot be consistently applied to quantum phenomena. All interpretational problems of quantum theory are surface symptoms of a deeper chronic issue regarding the nature of quantum reality. With a clearer understanding of this reality, these problems can disappear.

In over a century of quantum theory and its interpretations, a better understanding of its reality hasn't emerged. This is because physicists haven't been able to identify new intuitions using which the bizarre ideas can be demystified. Most physicists believe that intuitions about quantum theory cannot be identified from everyday experience because quantum theory pertains to sub-atomic particles, which don't have direct counterparts in everyday experience. Related to this view is the idea that quantum theory at larger scales eventually reduces to classical mechanics, when Planck's constant ħ becomes 0. Thus, the everyday world is presumably correctly described by classical theories while the atomic world is described by quantum theory. This selective application of quantum theory to atomic phenomena is at the root of our inability to look in the everyday world for novel ideas that can solve the quantum mysteries. The fact is that the quantum world *never actually becomes the classical world*, because Planck's constant ħ never actually becomes 0. The tremendous empirical successes of the theory must be taken to mean that the universe—even at everyday levels—must be described by quantum ideas. Everyday experiences can therefore be used to identify object concepts to demystify the quantum issues.

The main problem in interpreting quantum theory is the many kinds of conflicts that the theory presents with the classical worldview. Of course, we take the classical thinking for granted—it is not just how science started, but it is also the conceptual backdrop in terms of which we think of science itself. Classical notions about laws, object concepts, and theories have shaped our view of science. The quantum contrast with the classical worldview thus also includes a conflict with the current characterization of science itself.

Classical physics deals with point particles instead of discrete objects, uses deterministic theories to predict object states instead of uncertainty and probabilistic predictions, allows only local propagation of causes instead of non-local ones, uses only one kind of object notion at a time (wave or particle) instead of wave-particle duality, and all classical changes are reversible instead of the irreversible so-called collapse of the wavefunction. It would be therefore fair to say that the problem of quantum interpretation is its conflict with the established notions about matter and the science that grew from Newton's physics. The present chapter discusses the various ways in which quantum theory violates classical ideas.

Discreteness

Discreteness is the idea that quantum theory deals with discrete lumps of matter (or energy) and that matter isn't infinitely divisible into parts as in classical mechanics. While atomism was first advocated by Democritus in Greek times, there was no prior scientific evidence that there could be *limits* to breaking things down into smaller parts. Quantum theory not only reinforces the idea of discreteness (which has existed since the time of Democritus) but also defines the limits to how small each discrete entity can be.

But why is discreteness such a huge problem when ideas about atomism have existed for so long? The issue is that it puts the fundamental theory about matter at odds with the theory about space-time. Calculus (developed by Newton to make predictions about object motion) assumes that space and time are infinitely divisible. Continuity of space and time is necessary to define derivatives of space with respect to time at each point in space and time, which needs space and time to be infinitely divisible. Classical physics also assumes that matter is infinitely divisible to associate particle states with point positions in space and time. If space and time are infinitely divisible but matter is not, then the discrete states of material objects cannot be tied to point locations in space and time and since the classical law of motion depends on the ability to compute the second order derivatives involving space and time points, discreteness implies that these derivatives can't be computed. A fundamental limit in computing derivatives in calculus implies that the premise of the classical predictive framework is false.

Although quantum theory doesn't directly indicate discreteness in space and time, discreteness in observable, dynamical properties indirectly translates into the discreteness of space and time. This discreteness conflicts with the idea of motion that requires infinitesimal point locations in space and time. Discreteness of energy implies discreteness of time intervals. Discreteness of angular momentum implies discreteness of rotation angles. Discreteness of momentum implies discreteness of lengths. And discreteness of spin implies the discreteness of time directions. In so far as science is a theory about the motion of objects in space-time, the discreteness of material objects becomes the inability to accurately define object states in space-time and then make predictions about motion, rendering the classical scientific predictive framework ineffective.

Uncertainty

In the early part of the 19th century, mathematicians spent a significant amount of effort to show that space is infinitely divisible and that an infinity of points on a line can be traversed by an object in a finite amount of time. This was important to demonstrate that Newton's calculus had a firm mathematical grounding. It was also important to counteract Zeno's paradoxes[1] which claimed that if space and time were infinitely divisible, motion would be impossible. The mathematical notions about limits and continuity of space and time were invented to show that it was possible to compute the derivative of a trajectory at each point of the trajectory. The derivative of a trajectory represents momentum in Newton's physics. The existence of the derivative at each point on a line implies that position and momentum can simultaneously exist and are defined at each point in space. The particle therefore passes through each point in space and at each of these points it has a finite momentum. Quantum theory violates this basic principle derived from calculus.

Heisenberg's uncertainty principle asserts that both the position and momentum of a quantum particle cannot be simultaneously determined. If one of them is certain then the other is completely uncertain. This is mathematically expressed by the following equation where Δx denotes the uncertainty in a quantum object's position and Δp the uncertainty in momentum. Classically, this means that if a quantum object has a definite position then the derivative of its trajectory at that position does not exist. Hence, it is not possible to draw the trajectory of a quantum object in the classical sense.

$$\Delta x \cdot \Delta p \geq \frac{\hbar}{2}$$

The uncertainty principle similarly exists between energy and time, angular rotation and angular momentum. It implies that if time were known with certainty, the energy of the particle would be infinitely uncertain. Similarly, if the angle of the particle's state were known accurately, the angular momentum would be infinitely uncertain. There is some debate about whether this uncertainty pertains to the state itself—i.e. that the state is itself uncertain—or to our knowledge of that state (i.e. the measurement is uncertain). It is hard to justify either conclusion, because they have the same outcome—our determination of state becomes uncertain.

$$\Delta E \cdot \Delta t \geq \frac{\hbar}{2}$$
$$\Delta L \cdot \Delta \theta \geq \frac{\hbar}{2}$$

Another way to interpret the nature of uncertainty blames the act of measurement interaction. It says that a measuring device must exchange some energy with the measured system. This transfer of energy (and momentum) alters the state of the quantum system since by definition the quantum object is atomic and has a very small amount of energy and momentum as compared to the larger macroscopic measuring device. Thus, popular textbook accounts often conclude that the state of the quantum object is "disturbed" during the act of measuring. The reason I regard this view unsophisticated is that it tries to explain away a uniquely quantum principle in terms of classical notions about measurement. Why, for instance, would we need a quantum theory if the same inaccuracy in measurements applies to small classical particles as well? Also, since quantum theory is linear, small uncertainties in the current state should imply a small uncertainty in the next state, which is untrue.

Heisenberg's relations have an important significance: In classical physics, position and momentum form a complete set of properties (it was believed that if the position and momentum of all particles would be known at any time t, then the state of the entire universe could be predicted for any future time instance). The uncertainty in determining the complete set of properties simultaneously implies that we can never completely know the state of a system and hence can never predict its future state—a clear example of the indeterminism that quantum theory breeds.

Probability

Of course, quantum theory would not be a scientific theory if it did not offer any predictions whatsoever. While quantum theory does not offer deterministic predictions, it does make statistical or probabilistic predictions. Almost every interpretation of quantum theory spends their maximum effort in trying to explain the nature of this probability. Given that the goal of a scientific theory is to make accurate predictions, probability is the most obvious shortcoming in quantum theory. There is, however, significant debate amongst physicists and philosophers about the nature of this probability.

We can classify the different views about the nature of quantum probability into two broad categories, which I will call *epistemic* and *ontological* views about probability. The epistemic view says that the quantum object is in a definite state, but our methods of knowing the world through macroscopic instruments are inadequate to know the state accurately. Since we cannot know the world through anything but classical measurements, the limitations in knowing the world are here to stay and quantum theory basically reflects the fact that in our search for knowledge about the world, we have reached the limit of our classical style of measurements. Some might conclude from here that new theories and experiments may be needed. Others would insist that our theoretical framework is constrained by what we can observe through the five senses. Bohr notably held the latter view, concluding with the claim that quantum theory is incomplete and will always remain incomplete as a *human* theory of reality.

The ontological view about probability says that the quantum object is in fact in a "superposed" state. There is nothing wrong with our methods of knowing but with the nature of quantum reality itself, which does not exist in a definite state *a priori*. At the point of measurement, the superposed state "collapses" into a definite state. The nature of this collapse is probabilistic and hence predictions are probabilistic. Quantum theory cannot be deterministic because reality is in a superposed state and quantum predictions will forever remain probabilistic, in line with the nature of reality. Most quantum theorists today abide by this view. The majority focus of the research today is in trying to explain how the superposed state collapses into a definite classical state when a measurement is performed.

Einstein advocated a view different from both above notions about probability. His interpretation, which I will discuss later, says that quantum theory describes not an individual object but a collection of objects. He maintained that the statistical nature of quantum theory is an outcome of the fact that classical instruments can only measure a collection of quantum objects and not individual ones. He did not believe that the quantum state itself was superposed but that our instruments are unable to measure individual particle state. Quantum theory was therefore about our *knowledge* rather than the *reality* itself. Einstein did not consider this limitation to in the theory to persist for all times to come. He believed that we would one day construe a notion of the *individual* quantum object and the new theory would not be probabilistic because we would be able to measure the particle rather than the ensemble. Caricatures of Einstein's view sometimes fail to understand the profoundness of his view. These caricatures picture Einstein as a classical physicist who thought that nature must be classically deterministic and could just not come to terms with the newer quantum theoretic principles.

Wave-Particle Duality

Prior to the advent of quantum theory, light was described as a wave in classical electrodynamics and it could interfere with itself and other waves. When interference behaviors were observed with electrons (which would have been assumed to be particles) the first reaction was to believe that each quantum object was both a wave and a particle. This famously came to be known as the wave-particle duality. Subsequent formulations of quantum theory changed the picture significantly, although the terminology has unfortunately stuck. In quantum theory, the so-called wave is not a physical object, as was the case in classical physics. The wave in question is instead a complex valued function with no real-world interpretation. The quantum wave (called the wavefunction ψ) must interfere with itself to give real measurable values. These values however are not physical properties (in the sense of mass or charge or momentum) but a probability distribution over many possibilities.

The probability distribution avoids the need to think of the quantum object in contradictory terms as a wave and particle. The correct way to think is that current quantum physics cannot accurately predict measurement outcomes. When an ensemble of quantum objects is spread over a probability distribution, the result is identical to that of an interfering wave. But this is nothing more than a historical coincidence. If we reduce the intensity of light (thereby reducing the rate of quantum measurements), there are marked differences between the predictions of classical wave theory and those of quantum theory. Classical wave theory predicts that all locations in a probability distribution will be illuminated at once although with a dimmer intensity. Quantum theory instead predicts that only one location would be illuminated at any time with a fixed intensity. The quantum prediction has been experimentally validated, and this lends credence to the idea that a quantum object is always a particle. The so-called wave description pertains to a probability distribution of an ensemble of particles and does not represent any physical entity that truly exists in nature.

There is yet another reason why wave-particle duality is attributed to quantum objects. This is the photoelectric effect predicted by Einstein and experimentally proven by Andrews Millikan. After Planck had explained black-body radiation using energy quantization, few people believed in Planck's theory. Einstein was one of them. He extended Planck's hunch that energy is quantized by creating an experiment that would show that light is absorbed and emitted in packets. After Andrews Millikan experimentally proved this was true, it was established that light travels as a wave and is absorbed or emitted as a particle. There was a problem, of course, as we saw above, that in quantum theory we cannot speak of traveling objects, since we cannot measure position and momentum at once. In what way could we say that light as a quantum object could be traveling, and the *same* particle arrives at a destination when this is impossible in quantum theory? This problem has been largely ignored over the years due to measurements demonstrating that the time between emission and absorption using synchronized clocks reveals a constant speed of motion.

This constant speed of motion was previously based on the wave theory of light (again through a coincidence where a combination of electrical and magnetic waves predicted a speed identical to the measured speed of light, leading to the surmise that this coincidence must mean that light is an electromagnetic wave). With the discovery of the photoelectric effect, light now also had a particle behavior. So, light now had two behaviors—its traveling speed was explained by wave theories and its absorption and emission was explained based on particle theories. Thus, a photon was both particle and wave.

A third, more original, reason for the use of wave-particle duality is due to Louis de Broglie. He thought of quantum particles as looped stationary waves. A stationary wave is found in music instruments—e.g. in the vibrating string—which has both forward and backward components. Now you could warp the two ends of the string and join them together to form a circular stationary wave. This circular wave was indeed a wave, and yet it could be viewed as a particle. Accepting this interpretation would require the acknowledgement of two kinds of waves—one pertaining to the collection of particles and the other pertaining to the individual particle. This idea did not go down very well with the physics community. Thus, even though de Broglie coined the term 'wave-particle duality' his view about it has not stuck, although the term has continued to mean other things.

If photons are indeed particles, then the finite speed of light and indeed motion itself must be explained based on quantum theory. Such a quantum explanation doesn't exist today. Until we have a quantum explanation for the speed of light, the mysteries of wave-particle duality are not fully understood. After all, light is part of what quantum theory explains although the theory doesn't explain why the speed of light must be constant in all reference frames. The constant speed of light is also the basis of Einstein's relativity (both the special and the general variants), which contradicts quantum theory due to non-locality, and their unification remains one of the biggest outstanding problems of modern physics. From this angle too, a non-classical explanation of the finite speed of light is essential. In both cases, the speed of light is an empirically formed standard, rationalized based on classical electromagnetic theory. There is neither a relativistic nor a quantum justification for it today.

Non-Locality

Why are quantum and relativity theories contradictory? The answer lies in the changes that Einstein brought to classical physics by treating the finite speed of light as a *law* of nature. Every law, Einstein claimed, must be true in all frames of reference. If therefore, the speed of light is a *law* of nature then the speed of light is constant in all frames of reference. Of course, this is not true of other objects whose speeds are measured *relative* to the motion of the observer's frame. But it is true for the speed of light, which is constant across all frames. By treating the speed of light as a law, Einstein derived changes to the classical laws of motion. The modified laws state that as an object's speed increases, its mass also increases, such that the added energy will contribute towards increase in mass rather than velocity. Thus, an object can approach the speed of light but never actually reach it because after a point the object's mass is so large that any incremental addition of energy will only increase its speed to a very small extent. In quantum theory, the finite speed of light applies to the propagation of all fundamental forces, including gravitational force. This is hence not a unique property of light but a property of bosons or force particles. All forces must propagate at constant speed across all frames of reference, both inertial and non-inertial.

Einstein drew the conclusion that the speed of light is not just a *law* but it is also the highest possible speed for any material object. No material object can travel any faster. Instantaneous propagation of causal influences across large distances in space was presumed in

Newton's gravitational theory and Einstein showed that this should be false given a finite speed of light. Underlying this claim is also the idea that all causal influences involve transfer of energy. If all energy travels at a finite speed (light is energy) then all causes must also travel at a finite speed. By recognizing the speed of light as the maximum possible speed of energy, Einstein showed that causes can propagate at most with the speed of light, not instantaneously.

Paradoxically, Einstein also happened to be in the midst of the so-called EPR paradox[1] in quantum theory (the paper was co-authored by Podolsky and Rosen along with Einstein), which contradicts the conclusions earlier derived from the finitude of the speed of light. The EPR paradox is a result of the statistical view of quantum theory. Essentially, we take two quantum objects, whose combined quantum wavefunction is comprised of two states. These states could be up-spin and down-spin. Since there are two states and two particles, it is compelling to think that each particle is in one of the states. But quantum theory cannot predict which particle is in which state. Under the statistical view, we would think that particles are in a superposed state. Now, the EPR paradox allows the two particles to drift apart up to very large distances, such that the effect of the finite speed of light on causal propagations kicks in. If we were to perform measurements on these geographically separated objects, then according to quantum theory one particle will reveal an up-spin while the other shows a down-spin. Quantum theory cannot predict which location will produce which result. In the statistical interpretation, we would in principle assume that if these measurements were carried out multiple times, the results would be statistically distributed at each location over the two alternatives.

The EPR paradox now leads to the following problem. If a measurement on one of the particles has been performed, and the result showed an up-spin, then the outcome of the experiment can be predicted upfront from the fact that there are only two particles and two states in the wavefunction: the second particle will have down-spin. This fact poses a problem if we think of the quantum wavefunction ontologically. That is, if we believe that the quantum object is not in a definite state and the measurement puts it in a definite state through a "collapse" of the wavefunction, then it would seem that there is some faster than the speed of light communication transpiring between the two quantum objects to put the second object in a definite state per the result of the first measurement.

Two types of energy transfer schemes can be envisioned, and both contradict present theories. First, if energy propagates from the photon to the detector, EPR would violate relativity because there is a finite energy being transferred at an infinite speed. Second, if energy propagates from one detector to another, EPR would contradict quantum probabilities because these probabilities now arise due to classical communication between detectors. The second scheme would additionally contradict the relativistic principles.

The EPR paradox arises due to the assumption that a quantum is not in a definite state and measurements put the object in a definite state. Correlated outcomes between a pair of experiments violate the principle of locality (finite speed of light). Note that this would not be a problem if we treat quantum probabilities epistemically. That is, if probabilities refer to our ignorance of the system and not to the system itself, then non-locality arises from not knowing which particle has arrived at which location. If particles have a definite state although we are unaware of that state, then the non-locality experiment is no different than classical probability experiments done with two billiard balls. As an example, if there are two balls in a pipe, one black and one white, and we eject one ball from each end, then knowing the color of one ball at one end will also fix the color of the second ball at the other end. Classically, we don't know which ball comes out of which end and that is uncertain. But knowing the color of a ball at one end does not require subsequent communication of the color at the other end so the other ball can 'adjust' its color.

The problem of non-locality is therefore a problem with the ontological view of probability which entails that a quantum object is in an indefinite state and hence must communicate somehow the result of its measurement to the second particle in a pair to put the second particle in a definite state. The problem does not arise if quantum probability represents our ignorance. Of course, in the latter case, the situation begs the question why the theory is probabilistic and what could be done to fix the incompleteness.

Completing the theory in this case requires the prediction of which particle in a pair will arrive at which location, analogous to knowing which billiard ball will come out of which pipe end. Such a prediction would solve the problems of incompleteness, probability and non-locality without violating any of the theory's currently known tenets. Note how the ensemble view changes the problem from knowing the state of an individual particle (which appears to be an unsolvable problem about reality) to knowing which specific kind of particle arrives at a location. The ensemble view transforms quantum theory into a problem of not knowing object identities. In this case, the ensemble view says that there are individual particles, each of which exists in a definite state, and the identity of a particle is completely defined by its state. However, when we perform a measurement, we don't know which *specific object* the measurement is detecting. The problem pertains to predicting the *identity* of the next object instead of predicting the object's next *state*.

The inability to predict the next state of an object entails conflict with classical determinism. However, the inability to predict the next object being detected from an ensemble of objects does not imply such a problem. The problem instead pertains to how objects are organized within an ensemble such that successive measurements would reveal to us the internal structure of the ensemble of objects rather than the probabilistic state of an individual object.

Irreversibility

In present quantum theory the collapse of the wavefunction is the act of an observer outside the predictions of the theory. Between two collapses (measurements) the wavefunction evolves according to a deterministic equation, which is—in principle—reversible. The measurement event however isn't reversible. The irreversibility is tactically avoided within quantum theory by treating the act of measurement to be outside quantum physics. But there is no reason why the act of measurement must be outside physics. Is it not, after all, a physical phenomenon that should be explained by physics?

Theories about quantum measurement take one of two opposite views. The first view assumes that the measuring system is a classical object, which is in a definite state. The interaction between a definite state of the classical object and an indefinite state of the quantum object leads to a definite state in the classical object, which corresponds to the state of the quantum object at that point in time. In short, at the point of measurement, the quantum object is absorbed in the classical object, and hence becomes a classical object. This is a somewhat convoluted view of measurements but is by and large how most physicists currently think about the measurement problem. But why should we treat the measuring system as a classical object? Aren't measuring devices quantum objects too?

The second view treats both measuring and measured systems quantum mechanically. The problem is that the combined system of the measured and the measuring objects is a wavefunction which will remain in a superposed state, never collapsing into a definite state. The combination of two quantum objects thus never results in a measurement. An external agent—such as an observer's choice—must collapse the wavefunction into a definite state. John von Neumann suggested that consciousness "selects" an alternative, bringing definiteness to the probabilistic state through a "collapse" of possibility into a reality. This viewpoint leads to the mind-body problem, of how a non-material consciousness interacts with a material object. Quantum theory must choose between the mind-body problem and the classical-quantum boundary in case of measurements and most interpretations tend to choose the latter.

Some recent approaches have tried to mitigate the issues with the collapse hypothesis by replacing consciousness with physical effects. Incidentally, this coincides with the larger attempt in cognitive science to explain consciousness using a combination of physics, biology, neuroscience and computational theory. Thus, for example, Roger Penrose believes[2] that the quantum probabilities are collapsed by the effects of gravity and that gravitational force "selects" one out of many quantum alternatives. Similarly, Henry Stapp[3] attributes the classical world to the *Quantum Zeno Effect* in which a definite state is created by strong interference with an external electromagnetic field. None of these approaches are without their own problems. The main issue I find with these ideas is the belief that the everyday world is classical, because classical physics is not able to explain information and symbolic objects. Collapse approaches discard the most valuable insight quantum theory has to offer us to somehow fit nature into our classical sensibilities.

The classical-quantum boundary approach does not explain irreversibility. At best, we must assume that the classical world is itself irreversible in some way (the Second Law of Thermodynamics) and this irreversibility is reflected in quantum measurements. This pushes the explanation back into the classical world, where the particle theory is reversible, but the ensemble theory is irreversible. The use of mind-body boundaries provides a more intuitive explanation—that irreversibility is caused by a succession of conscious choices, and therefore conscious choices are irreversible—except that we still don't know how to solve the mind-body problem. We also don't know why conscious choices must be irreversible.

The question of irreversibility goes beyond measurement scenarios. There are cases when a measurement system is not involved, although probabilities are still involved. In such cases, one can clearly see irreversibility. An obvious example is radioactivity, where heavy atoms spontaneously decay into lighter atoms, although the decayed atoms don't spontaneously combine into original complex atoms. Per quantum theory, there is a half-life period of an atom which defines the probability of whether an atom will decay during a given time interval. However, the theory does not predict probabilities for decayed atoms regenerating into complex atoms. Since no measuring instrument is used in radioactivity, we must suppose that there is irreversibility in the combined quantum system itself due to which atoms spontaneously decay but don't spontaneously combine. The problem is now even more mysterious because probabilities don't pertain to measurements and radioactivity probabilities still apply regardless of observation.

3

Developing the Intuitions

We must be clear that when it comes to atoms, language can be used only as in poetry. The poet, too, is not nearly so concerned with describing facts as with creating images and establishing mental connections.

—*Niels Bohr*

Three Kinds of Possibility

To describe the world as meanings, we must distinguish between universals and individuals. A chair, for example, is the combination of a universal concept 'chair' and an individual, which makes it an individual instance of the universal concept. However, sometimes, this chair may be used as a table, in a specific context or relation between such instances of universal concepts. This leads us to a distinction between *ideas*, which are universal, and *meanings* which are contextual. The individual can represent a universal idea and a contextual meaning, so it is the bridge between the two.

This notion about the objects is commonly found in our everyday world when we refer to ourselves. For example, we use the world "I" in three different ways. First, we use "I" relationally. Common occurrences of this use are "I am a father", "I am an employee", "I am a citizen of this country", "I am the son or daughter of so and so", "I am married to so and so", etc. These are the ways in which we express our identity in relation to something or someone else and give ourselves meaning in relation to something else. It is like the postmodernist socially-constructed notion of the self. The post-modernist argues that the "self" is nothing other than the collection of all the relationships one bears to others. This idea about the self is not entertained in classical physics, because the identity of each object is supposed to be independent of the other objects. But this notion of identity exists in human relationships. If this idea is applied to matter, then material objects would have different identities in relation to different objects. This would radically change the notion of identity, because it would now be relationally defined.

Second, we use "I" materially and causally. Common examples of the material expression are "I am white", "I am black", "I am tall", "I am short", "I am fast", "I am intelligent", "I am fat", "I am thin", "I am a little slow today", etc. These are ways in which we describe ourselves as a body, which then has causal expressions such as "I am running", "I am reading", "I am talking", "I am writing", "I am sleeping", "I am working", etc. These are all the types of actions we perform.

Third, we use "I" emotionally. For example, we say, "I am happy", "I am jealous", "I am in love", "I am angry", "I am disliking this", "I am enjoying this", "I am desirous", etc. These are all the ways in which we describe our feelings, desires, likes and dislikes, and emotions.

Conscious experience is created when the three components combine. So, in every conscious experience, there is a relation to something, a cognition and action toward that thing, followed by an emotion about it. The order of these three things can be changed. Individually, before this combination, each of these three exist as a possibility. Thus, when I say that "I am a father", it doesn't mean that I am always acting like a father. My being a father is a potentiality. In fact, we have the potentiality of many relationships, such as father, son, employee, citizen, etc. But these potentialities are realized occasionally. If we take away all these interactions, then all these relationships become potentials. And these potentials become real one by one. Similarly, my body is also a potential. We call these potentials abilities. For example, I can read and write, but that doesn't mean that I'm always reading and writing. Similarly, I can eat food, or walk, but that doesn't mean I'm always eating or walking. So, the body is a collection of abilities which lie dormant and become active. When they become active, they need to be active in relation to something. For example, if I'm going to eat some food, then there must be some food. And once I eat the food, there will be either some feeling of satisfaction or dissatisfaction. So, what we call our body is not some material stuff. It is rather a collection of potentialities which become active when they contact other potentialities. Finally, my emotions are also potentials. Various types of emotions such as love, anger, jealousy, fear, happiness, sadness, etc. exist in me in potential form. Of course, I can feel some emotions more than others. But that doesn't mean that I am always feeling a kind of emotion. So, these emotions are potentials, and they constitute my individuality.

So, 'reality', has three parts—the individual, the universal, and the relational. However, this 'reality' is still a potential. Each of these potentials constitutes a meaning—although a different kind of meaning. And as meanings they are eternal and unchanging. The *combination* of individual, universal, and relational can however keep changing. Therefore, the idea 'man', the relation of a 'father', or the feeling of 'happiness' are universal. But a man feeling happiness in the relation of a father is not eternal. What we call change is therefore not the 'motion' of objects. It is rather the changing 'associations' between the individual, universal, and relational. That is, an eternally possible individual combines with an eternally possible universal and an eternally possible relation to produce an experience.

This intuitive description of our ordinary experience can be applied to the study of material objects, if we just remove the notion of 'experience' from it, and retain the components—individual, universal, and relational—which are treated materially. Each of these exists as a possibility; in current quantum theory, we have all three, although not clearly distinguished. For instance, the problem of probability exists because all the relations are not always realized. Unlike classical physics in which every object always interacts with every other object, in quantum theory, objects occasionally interact with each other—just like you do not always act as a father or mother although the potentiality of being a father or mother exists. The occasional exhibition of father or mother behavior is probability.

These relations are said to be 'collapsed' by a 'choice' (in the von Neumann interpretation) but the mechanism of 'choice' is not explained. It can be explained if the underlying reality was a *purpose*, which is experienced in living beings as emotion. The purpose can be objective while its experience as emotion is subjective. The purpose can be said to select a relation and instantiate an interaction.

Finally, once this instantiation occurs, one of the possibilities from the universal, or an *aspect* of that universal is realized, and these aspects are mutually entangled because they are part of a whole and bear a relation to it as parts. For instance, a macroscopic object has 'sides' like front and back, left and right, up and down, etc. They are parts of the whole, and if the whole exists, they are aspects of that whole, and therefore remain inseparable. This inseparability is the cause of the so-called property of 'quantum entanglement' due to which the parts of the whole do not behave as independent particles, as they would be if there wasn't a whole comprising of parts. When a relation is instantiated through a choice, one aspect of the whole is revealed, quite like we might observe a macroscopic object from the front or back, upside or downside, or left or right. They are like the faces of a dice, but one face turns up on observation as each face can only be seen via a relation to a measurement instrument.

The faces of a macroscopic object are discrete; they are not infinitesimal. Therefore, the particles on observation are also discrete. Finally, these faces have to be described as *symbols* which are like 'words' representing meanings. Their particle nature is that they are individuals, and their wave nature is that they are vibrations which denote meanings; the meaning is the relation to the whole. For example, in the case of a chair, the different parts can be described semantically as legs, seat, and backrest. As independent parts, they are all pieces of wood. But within a chair, they have meaning.

All problems of quantum theory—discreteness, probability, wave-particle duality, non-locality or entanglement—are amenable to an intuitive understanding of the everyday process of perception involving macroscopic objects. This understanding can be extended to atomic theory as well; the hurdle isn't the that the intuitions about atoms are absent; the hurdle is the classical physical description of this world which hinders a clear understanding because the ideas underlying this classical description are patently false. The idea of infinitesimal particles, the independence of these particles, the reduction of the whole to the parts, the continuous interaction between the parts, and the reversibility of the mathematical formalism that results from these assumptions prevents a grasp of quantum problems unless we discard all these assumptions.

The main insight I would argue for is that there are three categories of meaning—the individual purpose, the universal idea, and the contextual relation—which combine to create a quantum object. The individual purpose is responsible for the collapse, the universal idea for the entanglement, and the contextual relation for the probabilities. The quantum is discrete as a symbol, which has individual, contextual, and universal properties. Therefore, it is not possible to describe this world using classical concepts because there are contextual and universal properties in the symbol. We cannot reject the individuality because quantum particles are detected as individuals. We must therefore find a way in which to reconcile the individuality with contextuality and universality, and then describe each of them as possibilities rather than an observable reality.

Quantum vs. Classical Space

The rejection of the classical worldview is fine, but it doesn't explain why classical physics works so well. To explain this problem, we need to recognize the existence of the classical world as different from the quantum world, and then bridge the two worlds.

I will now introduce a radical idea in which quantum and classical worlds are present in different *spaces*. The quantum world is reality, but it is never observed because it exists as a possibility. The observation converts the quantum world of possibility into a classical world of phenomena. This phenomenal world is produced due to an interaction between the objects in the quantum world.

Let's illustrate this with the example of *distance*. While writing this book, I'm facing my computer, and I believe that I have proximity to it. Semantically, however, I and the computer are radically different meanings, and I am far apart from the computer in a semantic sense. My sense of proximity to the computer is the classical world that arises due to an interaction between far-off semantic objects. That is, when we interact strongly, I appear to be close to the computer, even though we are semantically quite different. Similarly, when we stop interacting, we seem far apart. You can imagine this further through a table and a chair. These are semantically different, so they are far apart as concepts. But they can be physically brought closer if proximity were defined as a stronger interaction while distance was defined as a weaker interaction between them.

This idea is consistent with classical physics, although the cause-effect relation is inverted. In classical gravitation, for example, the distance between particles is taken as a fact, and the strength of the gravitational force is the inverse square law of this distance. That means, proximity is associated with a strong interaction and distance as a weaker interaction. I'm only arguing for the reversal of this relation: I'm claiming that instead of treating the distance as the cause of the force between them, we can treat is the *effect* of forces. That is, the distance we observe classically is not real; it is produced due to the interaction, which we call the classical physical space.

The reality is the quantum objects which are not classically observed, but they have interactions between them. When two quantum objects interact strongly, a sense of classical proximity is created between them. When they interact weakly, a sense of classical distance is created between them. Therefore, 'force' or interaction between particles is the cause of classical distance. Once this classical sense of proximity and distance is rejected as a reality, then we can speak about the quantum world and its interactions as the real world. This world must also exist in a space—which is a quantum rather than a classical space—i.e. it is *real* rather than produced from an interaction. As a world of possibilities, it is always fixed; i.e. the objects in it are not moving. However, the interaction between these quantum objects then creates a classical world.

Change in classical physics is described as motion. But change in the quantum sense must be described differently. As possibilities, all individuals, universals, and relations are possible and hence 'eternal'. However, the *combination* of an individual, an universal, and a relation is temporary. Therefore, change in the quantum sense is the changing combination of the three categories of meaning which produces quantum objects, and the changing force of interaction between the objects which produces the phenomenal or classical world. Classical motion isn't therefore simply a product of forces between classical particles.

It is rather the product of forces between quantum particles, which themselves are comprised of three aspects.

Even though we observe it empirically, the classical world—i.e. the position or distance and momentum or motion—are illusions. The distance between objects is an outcome of the strength of interaction between quantum objects, and the motion of these objects must be attributed either to the change in the combination of the three types of meaning, or to the changing strength of interaction. Classical motion and position therefore must be explained based on a new kind of reality (the three components) and causality (the interaction between the combination of these components).

In this regard, we must remember that the dynamical variables in quantum theory—such as position, momentum, angle, angular momentum, time, energy, time direction, and spin—have little to do with their classical counterparts. They are based on the quantum eigenfunction rather than a classical particle. The eigenfunction is itself a discrete position, has wave-particle duality, entanglement, probability, and its collapse sequences are irreversible. So, despite using the language of classical physics—i.e. the same dynamical properties—the significance of these properties is different in quantum mechanics. We should therefore treat these properties as representations of meaning, rather than as physical traits.

Why Physicalism is Inadequate

A key requirement for physical theories is that they be *complete*. A complete theory describes all aspects of an object in terms of some of its properties. Once these properties have been measured, then all the object's behaviors can be predicted in terms of those properties. Classical physics and current quantum theory describe matter in terms of physical properties and the central problem of quantum theory is that not all the object's behaviors can be accounted for in terms of its physical properties. Can we now assume the existence of some additional physical properties that complete quantum theory? There are three main hurdles in making this assumption. First is the ideological bias where we think that physical properties exhaust all the scientifically measurable traits of an object. Second is the pragmatic difficulty in identifying such traits for quantum objects. Third is the apparent contradiction with Bell's Theorem which shows that any theory of quantum phenomena that postulates "hidden variables" will contradict the observed non-locality. Here, I will try to address these hurdles through the notion of information.

Information is an objective trait of symbols, and a symbol indicates more things than we would know only from its physical properties. This addresses the first issue about physical properties being adequate for a description of nature. If nature is symbolic, then physical properties are not enough. The difference between symbols and classical objects is given by concepts and programs and these semantic properties can help us overcome the pragmatic difficulty in theorizing about the missing properties of quantum objects. This answers the second objection about how we identify new properties for explaining quantum phenomena in a new empirical way. Thirdly, information requires no hidden variables because the meaning is decoded by interpreting tokens of meaning in terms of a language, although current *physics* does not allow the possibility that nature has a language. Gleaning information from matter requires a new way of looking at currently well-known empirical facts—i.e. interpreting physical objects as tokens of meaning. Information therefore is not hidden; it is both objective and empirical and this addresses the concerns around Bell's Theorem. By using information as a new category of measurable traits about symbols, we can address the quantum problem, without adding any new observables to the theory. Information, however, requires interpreting observables in a new semantic rather than physical manner.

The attempt to identify new properties to address the quantum problem isn't new. For a while after the birth of quantum theory, many physicists believed that the quantum theory's description could be completed by some additional 'hidden' properties. These attempts at completing quantum theory tried to bring to the theory the kind of determinism that existed in classical physics. Since classical physics involved local causality, these theories came to be viewed as *local* hidden-variable theories. In 1964, John Bell proved the impossibility of using local hidden variables to complete the theory in one specific respect—non-locality—because non-locality explicitly violates the local pictures of causality used in classical physics. Thereafter, every attempt to complete the theory by postulating extra properties is taken to deny Bell's Theorem. But this isn't necessarily true. Locality depends on conservation laws; material transactions must be local because matter and energy are conserved. However, quantum non-locality does not involve a transfer of energy or mass. And since every communication in the classical sense requires an energy transfer, quantum non-locality is not a communication in the classical sense. The hidden variables in a classical sense can't be used because we can't assume any energy or mass transfer, to communicate between entangled particles.

Quantum non-locality can be understood by postulating that quantum particles are entangled due to an informational relation and not due to energy transfer. For instance, the meaning of the symbol 'hot' is defined in opposition to another symbol 'cold'. Neither symbol has an independent meaning; they are meaningful by their relation. Each symbol is already in a definite state, and not in a superposed state. Therefore, a "collapse" is not necessary to put them into a definite state. They are entangled because of their semantic relation. Non-locality, now, is a function of the distinction between symbols, rather than of matter or energy transacting between them.

Of course, there is still the need to overcome probability and explain indeterminism, uncertainty, irreversibility and duality. As it turns out, the notion of information also explains the other unintuitive aspects of the theory. It also turns out that information makes new empirical predictions, because probability represents missing information in a semantic proposition. The postulate of information helps us understand why present quantum theory, which reduces information to physical states, must be probabilistic, uncertain and dual, because notions of classical physics describe material particles but not symbols. Probability and uncertainty aren't features of atomic reality although they are features of the present theory and can be overcome in an informational treatment. Non-locality, uncertainty, probability and duality represent a non-classical behavior of information within a classical formalism.

What is information? How is information different from physical properties? Can information not be reduced to classical properties? While a detailed analysis of this problem will come later in the book, let me try to answer these questions using examples of information in ordinary sentences. This analysis shows that the information content in a sentence goes beyond what can be known via physical properties. Information cannot be reduced to physical properties and a new view of nature must be postulated to account for matter that represents information. A mathematical theory of information is however left to a later effort, because, as I will show shortly, a semantic theory of nature requires a semantic mathematics, and current mathematics does not handle the problems of meaning. Let's presently turn to information in ordinary sentences.

a. Mary has two bachelor children, John and Vincent.

b. Mary has two children. Their names are John and Vincent. John and Vincent are presently bachelors.

c. The battle of Mahabharata was fought among estranged first cousins almost five thousand years ago.

d. Complex interest is computed by calculating the simple interest upon the monthly reducing balance.

For the sake of the following discussion, let us assume a simple definition of physical complexity. This definition is not crucial to the main argument here, and other definitions may be used equally well. But this definition simplifies the crux of the argument, without complex mathematical notation. So, let's assume that the physical complexity of a statement is the total number of characters in the sentence. This definition is simple, but it captures the essence of physical measurements that depict objects in terms of quantities.

Statements B, C and D above have 101 characters each while statement A has 52 characters. The physical complexity of A is therefore nearly half of the complexity in B, C and D. However, A conveys the same meaning as statement B. Sentences A and B have different physical complexities although they have the same semantic complexity.

The differing measures of physical complexity will not tell us that statements A and B are identical semantically. Similarly, statements B and C have the same physical complexity, but C has a more complex meaning than B. At this point, we can say this because it is hard to compress C to a shorter sentence, although B can easily be compressed; A is the compressed version of B. The physical measure of complexity between sentences B and C does not tell us which statement is semantically more complex. Statement D is comprised of the same number of letters as B and C, but it provides a procedure for *calculating* complex interest. Statements A, B and C in contrast only describe but not calculate. D has computational complexity apart from semantic and physical complexity.

The information encoded in the above sentences must be decoded in a way different from just physically analyzing the symbols. Postulating additional physical properties in the sentence would not help us determine the meaning. We instead require a different method for *interpreting* a sentence to get the meaning. A physically complex sentence could be semantically simple, and a semantically complex statement could be physically simple. Physical properties don't therefore exhaust the semantic content, although we can assume that, on average, more characters would be needed to represent complex meanings as compared to simpler meanings.

The incompleteness of quantum theory can be because we haven't identified all the ways in which reality can be interpreted. Interpretations here aren't just philosophical views about reality. Rather, interpretations give a measure of complexity in matter which is useful scientifically. When we treat physical objects as symbols, there are three ways to interpret these objects—physically, semantically and computationally. Thus, a symbol can represent a thing, a meaning and a formula. A statement could have various degrees of physical, semantic and computational complexities, and the knowledge of the object is not exhausted by the measurement of physical properties. Thus far, an object's semantic and computational complexities haven't been considered relevant. So, if the behavior of an object depends on semantic and computational properties, then a theory that predicts based solely on physical properties would be incomplete. We can speculate that quantum theory, which uses only physical properties, would be incomplete if in fact it is dealing with a reality that has semantic and computational attributes.

What kinds of phenomena require information? The simplest example of information is communication. Communication is public and can be measured, as opposed to mental thoughts, which are private. The analysis of meaning centered on communication can avoid the issues of mental privacy. Objects in everyday experience are treated as *communicative* agents, not just as things. Such objects include books, pictures, and programs, and they have both physical and informational properties. Physical descriptions of reality do not exhaust all aspects of an object relevant to their *communication.* When matter is a communicative medium, then science needs to deal with additional aspects of matter than in classical physics. Additional properties of communicative objects don't have to be limited to macroscopic objects. Atomic objects too can be communicative because information can be encoded in them. SI is based on the premise that quantum theory pertains to symbols of meaning.

Books, pictures, and music are communicative objects. They have physical properties, but the information they express goes beyond these properties. The content in a book or a picture can be imprinted on paper, magnetic media or electronic circuitry. While these methods of imprinting differ in their physical complexity, all these representations are identical in terms of their information. Current science can describe the physical complexity in objects but not the kind of information they are representing. This is because science has thus far studied the world with a mind-body divide: the information about the world is in the mind, not in the world.

In the semantic view, information is a material property when matter represents ideas, theories, descriptions and programs. It is imperative, therefore, to bring information into the world and into the scientific theories about reality. Information, however, isn't a property of matter in current science and adding information to science requires adjustments to physical theories. Quantum theory hints towards how these kinds of adjustments can be made.

We can think of theoretical adjustments in two ways. One, we can imagine that information requires entirely new scientific properties that have no physical realization. Two, information could be represented by some physical attributes that are already well-known. While both approaches are equally useful as far as the study of meaning is concerned, the second approach is obviously more fruitful if this study must be pursued within physics. SI pursues the latter alternative. In SI, the physical properties of quantum objects themselves represent meanings. The main task of interpretation is figuring out which physical property corresponds to which semantic aspect of a symbol. This is what this book will attempt to achieve.

Reinterpreting Reality

We live in an era of mind-body dualism. René Descartes, who proposed the dualism wanted to give science the freedom to pursue its investigations without being encumbered by religious necessities imposed by soul, God and afterlife. To achieve that, he claimed that religion pertains to the mind and science to matter. Since these are distinct realities, science can pursue its objectives independent of religion. This division has proved useful as a scientific methodology. But the dualistic view of reality creates problems that we haven't dealt with so far. In the Cartesian view, the knowledge of the world is not in the world. Knowledge is a property that belongs to the mind. That puts the meanings of symbolic objects such as books and pictures into the mind. It is hard to accept that the meaning in a book is entirely a construction of the mind and there is nothing objectively meaningful in the book itself. And yet, this is precisely the conclusion that we are led to under mind-body dualism. This conclusion has been embraced both by modern science and philosophy.

One consequence of this dualism is that knowledge of the world does not play any role in the behavior of matter. So, whether you know about gravitation does not affect the motion of planets. But this also entails that your vote for Conservatives or Liberals does not depend on whether you know what these terms mean! Nature is, in a sense, "blind" to knowledge. Knowing the world does not change anything in the world, because knowledge only represents the facts about matter outside of matter—into the mind. Given the consequences of this extreme view on the human world (we do believe that our knowledge makes a difference to our lives) there has been a scientific need to explain the interaction between mind and matter. But how can two different realities interact? The point at which the mind interacts with matter, the mind becomes matter.

The solution to the mind-body problem in SI is that matter itself represents information, but it must be described differently. First, nothing is meaningless; everything we can observe is a token of meaning. These tokens are created by the combination of two things: a concept or a universal, and an individual or object. Due to the concept, each symbol as a universal meaning. However, this universal meaning gets further nuanced or modified through relations to other symbols. Since each symbol has a universal meaning, we can apply concepts to each object—e.g. a macroscopic object can be called a table or a chair. Furthermore, since meaning is modified through relations, a table may in some context be described as a chair.

The meanings 'table' and 'chair' are not Platonic independent ideas. Rather, these two are related to a more abstract notion of 'furniture', which produces a hierarchy of meanings. Therefore, even the concepts get their meanings in relation to other concepts, and these meanings are further modified in relation to other symbols. In a simple sense, we can say that meanings always arise within a collection of particles, and collections therefore compel us to consider meanings as important causal properties. If we studied the world as independent particles, we would fail to see these causal properties, and the description would be an approximation.

Since meaning arises through relations between particles, we must treat the locations and directions in space-time semantically. In classical physics, any particle can be at any location. Quantum theory fixes the locations of quantum particles, in relation to other particles in the ensemble. The location of a particle is not a possessed property of the object (like mass and charge), although it is a fact about the object. We can say that the location is a *contextual* property of the object and not a physical property. If this location could be given a semantic interpretation, then it would be possible to formulate a theory of how meanings can be encoded in material objects.

Location is a property of space. So, while Descartes thought that meanings cannot be encoded in *res extensa*, or objects in space, that view may have been wrong if positions of atomic objects can be given meaning. Similarly, quantum theory also fixes the directions of objects. Again, directions are not possessed physical properties of objects although they are empirical properties. Direction is also a property associated with space. Like locations can be used to denote meanings, directions can also be used to represent meanings. Distance and direction establish a relation to another symbol—distance indicates the value of a certain type, and the direction indicates the parity of that value of that type. For example, consider the property of heat, which represents an axis. It has many possible values such as very hot, hot, warm, etc. Then it has a direction which represents the opposite types such as very cold, cold, and cool. The combination of the axis, value, and direction creates types. However, we must note that the direction is defined on an axis in relation to the types in the opposite direction. Just like +5 and -5 are opposites on the same axis, similarly, very cold and very hot are opposites. They are not *quantitative* opposites but *qualitative* oppositions.

SI is the view that distance and direction are not possessed physical features which 'belong' to the quantum objects. They are rather fixed by the contextual collection of particles. Meanings too are properties of collections and not of individual particles. This isomorphism between meanings and space helps us map the two. We can say that quantum objects acquire meaning when located in space. These features make quantum objects *symbols*. When such symbols are sequenced in time, then we get the evolution of meaning. When the sequence has a time direction[1] (succession in addition to order), then it can represent progress or regress in the knowledge.

Location in space is associated with linear momentum. Direction in space is associated with angular momentum. Location in time is associated with energy. And direction in time can be associated with spin (I will elaborate on this point later). Momentum, angular momentum, energy, spin, position, space-direction, time and time-direction constitute the fundamental dynamical properties of all quantum objects. They are dynamical because they can change. If position, time, space-direction and time-direction can be given a semantic interpretation, then momentum, angular momentum, energy, spin can also be given a semantic interpretation. These two properties can represent descriptive and programmatic meanings, respectively. Therefore, the position, time, space and time directions will represent what an object is conceptually, while momentum, angular momentum, energy, and spin will denote what it can do. This 'doing' is also conceptual. For instance, in everyday language we associate a conceptual object like a knife with acts of cutting. The term 'knife' is descriptive, while the term 'cutting' is programmatic. All dynamical properties of objects can therefore be semantic.

Note that all the above dynamical properties are *contextual* properties in quantum theory because they are fixed for a quantum object in an ensemble provided the total energy of the ensemble is not changing[2]. In current quantum theory, these properties are derived from *possessed* properties of particles such as their mass and charge, like classical physics. These possessed properties, however, do not play a causal role in quantum theory as they did in classical physics. The possessed properties are only used to compute the dynamical properties, and all causal behaviors are associated with dynamical properties. We could therefore invert the quantum formalism and derive the possessed properties from the dynamical properties. In current quantum theory, the possessed properties are fundamental, and all other dynamical properties are byproducts of the possessed properties. This view of the formalism extends the particle-based thinking in classical physics and underpins the Standard Model of particle physics. But it is also possible to treat the dynamical properties as fundamental and compute the possessed properties from them. If the dynamical properties represent meanings, this approach would imply that the possessed properties are outcomes of encoding meanings in spacetime. The particles of the Standard Model in particle physics would now represent elementary *tokens* that can encode various types of meanings.

Which of these viewpoints is better? If we think of nature as particles, then the possessed properties are more fundamental. However, if we think of nature as meanings then the dynamical properties are more fundamental. The fact that a quantum object with a given possessed property (e.g., fixed mass of an electron) can be given many dynamical properties (e.g., position) lends itself to a simple interpretation where a given amount of matter (e.g., mass) can be used to encode a variety of different meanings (e.g., position). If we begin with the idea that nature is physical particles with possessed properties then it is difficult to understand how these particles acquire contextual properties, unless we note that the dynamical properties of quantum objects are themselves contextual properties. But we can also begin with the idea that meanings (and therefore dynamical properties) are fundamental and these meanings are used to create particles with possessed properties. The latter viewpoint will imply that meanings are logically prior to matter, and they are used to create possessed properties.

If we only measure the possessed properties, then we cannot know the meanings in the objects. However, together with dynamical properties it is not difficult to see that the quantum object could represent a symbol of meaning where the meanings are given by dynamical properties. The radical departure from classical physics now is that the dynamical properties themselves are meanings.

The possessed properties in the quantum allow us to think of the quantum as a *thing*. The dynamical properties however allow us to think of the quantum as *meaning*—because they are contextual. The redefinition of reality in quantum theory is that what we call matter is possessed properties like mass and charge. But these properties existed in classical physics as well. Similarly, the dynamical properties too existed in classical physics. The difference is that the possessed properties now exist in quantum theory in the same way as they existed in classical physics, but the dynamical properties are different as they are contextually defined in an ensemble of particles. A classical system does not fix the dynamical properties even for a closed system. Rather, dynamical properties in a classical ensemble continuously change (e.g., we can envision a collection of classical particles that are colliding with each other therefore changing their position and momentum). Therefore, if we did associate meanings with the classical dynamical properties, then there would be no encoded meaning as it would constantly evolve. The mathematics of classical physics will also imply that meanings can be infinitesimally small and each meaning in a symbol is independent of the meanings in other symbols. Associating meanings with classical dynamical properties will therefore violate everyday intuitions about meanings. The fact that classical intuitions are violated in quantum theory can therefore mean that it can be a satisfactory theory about meanings although this would not have been possible in classical physics.

This raises a fundamental question about the nature of dynamical properties in classical and quantum theories. Do the words 'position' and 'momentum' have the same meaning in classical and quantum theories? I don't believe so. There are many reasons for this. Position and momentum cannot be measured at once in quantum theory while they could be measured at once in classical physics. Position and momentum were two different state variables in classical physics, but they are *complementary representations* of the *same* state in quantum theory. Position and momentum implied a state of motion in classical physics and they imply a stationary state in quantum theory[3]. By no stretch of imagination can we thus treat these two words as denoting the same ideas in the two theories.

Classical dynamical variables are a non-semantic approximation of the quantum semantic world. In this approximation, we measure the weight, size, and number of pages in a book, but not the book's meaning. This approximation entails that reality is material objects whose meanings are in the mind. Quantum dynamical variables can, however, be interpreted to denote a semantic world in which the dynamical properties of the objects themselves denote meaning.

The Role of Meaning in Mathematics

Information is just numbers, but with one exception. Currently, we treat all numbers as quantities. Information content is, however, conceptual and involves *types* which are perceived as meanings. Descriptions and programs require two different ways of representing types because they indicate two kinds of meanings. The meaning of a description is a concept and the name of the object to which the concept is applied. The meaning of a prescription is a program and the purpose towards which the program is used. To treat information as numbers, therefore, the numbers should be able to represent types— descriptive and prescriptive. Numbers already represent quantities, so a new number theory that treats numbers as names, concepts, programs and purposes is needed. This topic is central to SI, and I believe, to the future development of quantum theory. To illustrate the relation between quantum theory, mathematics and computing, I will take a short detour into the role number interpretations have played in mathematics and computing theory, and the problems that arise from these interpretations.

Number interpretations are at the root of two important issues in mathematics: Gödel's Incompleteness Theorem in number theory and Turing's Halting Problem in computing theory. Both these theorems have their roots in the ability to interpret numbers in ways whose meanings are well understood in ordinary language but can't be expressed in mathematics. Gödel's Incompleteness Theorem interprets a number as a name for a statement and as knowledge of the statement's meaning. We intuitively know how to distinguish between names and meanings in ordinary language, but we cannot do so in mathematics when both names and meanings are expressed as numbers. This is because names and meanings are separate *categories* in ordinary language, but they become indistinguishable when both are expressed as numbers. Gödel's incompleteness therefore arises because of our inability to distinguish between names and meanings in mathematics. Turing similarly interpreted a number as instructions and descriptions of a program. In everyday language, the program instructions denote actions while the program description represents the problem that the actions solve. A given set of actions can solve many different problems, and a given problem can be solved by many different instructions. Thus, we know how to distinguish instructions and descriptions in ordinary language, but there is no way to distinguish between them in mathematics when both are expressed as numbers. Gödel's and Turing's theorems indicate the need for understanding numbers in multiple ways. But a failure to implement these distinctions leads to the view that mathematics and computing are incomplete. I treat this subject matter more extensively in my book *Gödel's Mistake*.

The problems of incompleteness can be solved if we can distinguish between the various meanings of a number in mathematics and computing theory. Through these distinctions, we can relate the incompleteness in mathematics and computing theory directly with the incompleteness in quantum theory. Essentially, both problems arise from the need to view symbols as capable of representing names, concepts, programs and purposes. In mathematics, numbers are treated as purely syntactic tokens, shorn of their meaning and reference. Meanings enter mathematics when ordinary language categories are brought into mathematics. These categories *interpret* numbers as names, concepts, programs and purposes, creating problems of incompleteness and incomputability. If mathematics can incorporate meanings besides syntactical tokens, then it could also describe quantum objects as symbols. The short answer to 'What is information?' is therefore 'numbers' if we can treat the numbers as meaningful entities besides tokens. This implies that quantum objects can be modeled using mathematics provided numbers could be interpreted in multiple distinct ways.

Gödel's Theorem

Gödel proved the incompleteness of mathematics by constructing statements about other statements. Such a statement represents *knowledge* of the first statement; we should distinguish between the two as referring and referred statements. The referring statement makes claims about the referred statement and these claims can, in principle, be false. For instance, if we fail to understand the meaning of the referred statement, we can make a false claim about that statement. The falsity of the claims in the referring statement however does not falsify the referred statement because these are two different statements. This basic distinction between knowledge and reality however does not exist in the case of numbers. This is because a number can alternately be treated as a thing, as a name and as a concept. All these interpretations of a number are possible in number theory, but they cannot be distinguished.

By using an ingenious trick—in which a number is treated as a thing, the thing is given a name, and finally the name is interpreted as a concept—Gödel eliminated the distinction. The denial of knowledge leads to a self-contradiction and Gödel concluded that if mathematics is complete then it will have some contradictory statements. If, however, mathematics is consistent then it is not complete, and some true statements will not be provable. This conclusion is called Gödel's Incompleteness Theorem and it is one of the most stunning, perplexing results in modern mathematics[4].

Gödel starts by showing that all statements can be mapped to numbers. He then takes two kinds of statements—referring and referred statements and maps them to numbers. In general, we would imagine that referring and referred statements must map to different numbers, because they are different and uniqueness demands that we map statements to numbers one-to-one. This is how the proof starts and two statements can be as follows.

Statement Q: Statement P is Not Provable.

Here Q is a statement about another statement P. Q may be false or true, and at this time, unless we prove P independently, we cannot say whether Q is true or false. But Gödel did not worry about trying to prove the statement P. Instead, he showed that it is possible to tweak the numerical mappings such that the referring statement refers to itself, and Q is equated with P. Thus, by changing numbers it is possible to construct the following statement.

Statement P: Statement P is Not Provable.

In this statement, if P is provable, then P is false. If, however, P is not provable then it is true. This gives us two choices. Either mathematics has some statements that are false but can be proved. Or, there are statements that are true but cannot be proved. The former implies that mathematics is inconsistent and the latter that mathematics is incomplete. Mathematicians will prefer an incomplete theory to an inconsistent theory and Gödel's theorem is therefore called the Incompleteness Theorem. This hinges on the ability to map a statement and a statement about that statement to the same number. In everyday parlance, this means the ability to interpret a number both as reality and knowledge about reality.

Gödel's theorem represents the problem that physical properties in a symbol may contradict the meaning denoted by the symbol. We decipher the meaning in the symbol based on its physical properties (e.g., the symbol's shape). In that respect, the symbol's meaning is a reinterpretation of the symbol's properties. So, how can these physical properties contradict the meaning in it? The short answer is that Gödel computes names of sentences without considering the *meaning* of the sentence. In effect, Gödel takes a proposition "married man" and maps it to a name Mr. Bachelor, then interprets Mr. Bachelor as someone who must be a bachelor and arrives at the contradiction that a married man is a bachelor. This problem can be fixed by distinguishing between the name and the meaning (i.e. introducing grammatical categories in mathematics) or by ensuring that names are aligned with meanings. In the latter scheme, it would be impossible to map the idea "married man" to the name Mr. Bachelor. Indeed, it should only be possible to map the idea of a "married man" to the name Mr. Married Man. Such a scheme can be pursued in the case of concepts, but when we distinguish between universals and individuals, then only the former scheme is useful. That is, the name by which a concept is called must be identical to the meaning of the concept, but the name by which a person to whom the concept applies can be different from that concept, provided we can distinguish between the individual and the conceptual naming.

In the case of concepts, the solution to Gödel's paradox suggests a naming scheme where names accurately reflect meanings. Let's call this scheme *semantic naming*. In this scheme, the name 'red rose' must mean the concept 'red rose'. However, an individual rose that is red, can be called by another name, if we can distinguish between the universal and the individual. For mathematics, restricted to deal with concepts, semantic naming suffices. However, to deal with the application of mathematics to the real world, the distinction between conceptual and individual naming becomes necessary. In any case, if names and meanings are consistent, then contradictions can never arise. In current mathematics, neither the adaptation of the name to the meaning, nor the separation of the naming and meaning is possible. Therefore, mathematics must have contradictions.

In classical physics, any particle can be at any location. So, if the location denotes the name of the particle, but the properties of the particle are contradictory to the location, then we can create a contradiction by naming the particle by its location and then deriving its

properties based on that location. This problem doesn't arise in physics because we distinguish between the location and other properties
(such as the mass and charge of the particle). Essentially, we attribute
two separate properties to a particle to avoid the contradiction, when
ideally, we must say that the name is the particle's *identity* or individuality, and the meaning is the properties associated with it. Potentially,
these properties (e.g. mass and charge) could exist in other particles
as well; so, they are not unique to the particle. The location is, however,
unique to the particle (other particles cannot be at the same location
at the same instance in time). Therefore, if naming is the space-time
location, and the meaning is the other properties, there would be no
contradiction.

This problem is better understood in the context of quantum theory
where each location is given by an eigenfunction, which represents the
particle's identity. The 'location' is not a point, but a function extended
in space and time (along with direction in space and time). And its
properties are the energy, momentum, angular momentum, and spin.
The eigenfunction, however, completely determines the other properties, which means that knowing the location and direction in space
and time fully specifies the other properties. This is a case in which
naming defines meaning, and one of them is determined by the spec-
ification of the other. This is the type of reality which we described as
being 'conceptual' above. In contrast, the classical reality is 'physical'
in nature. In the physical reality, any name can indicate any meaning;
as a result, the position doesn't determine the momentum, or vice
versa. In the conceptual reality, the naming determines the meaning.
Hence, we should treat the location and direction of a quantum particle as representing a meaning.

These locations or names existed in classical physics too, but there
we could not equate them with meanings. Classical physics is therefore
a world in which a 'red rose' can be called a 'yellow rose', and we must
distinguish between naming and meaning, although mathematics is
incapable of making this distinction, so physics adopts two separate
properties—position and momentum—to specify the particle's state.
In quantum theory, however, knowing the position state is enough to
know the momentum state. As a result, the naming and meaning are
identical. This can be said to form the basis of the quantum-classical

difference: in the classical world, naming and meaning are unrelated, but in the quantum world they are. As a result, we must treat the names as representing meanings.

By doing coordinate transforms, you can change the naming of an object, but the meaning remains unchanged. If, therefore, naming and meaning are identical, we cannot do coordinate transforms that change the names. It follows that to consistently hold a semantic viewpoint, we need to acknowledge a universal reference frame. This reference frame corresponds to a *language* of encoding meanings. By changing frames, we change the *words* by which meanings are represented. We change the language in which nature must be described. Can we do such a transform? Certainly, we can adopt different languages to describe nature, but those won't change the meaning. But, by changing the language we will make that knowledge incommunicable to those who use a different language. Science—through describing the truth—will become relative to the person and language describing it. As a result, we must use a universal reference or language that uniquely maps meanings to names, and the best scheme is one that employs the same *words* for describing the two.

When we encode meanings in physical states, the biggest problem is how to find whether those meanings are true. The facts intended in every statement can be potentially true, although they may not *refer* to the correct objects. This reference can be adjusted by renaming objects through a coordinate transform. For instance, if a statement says "X is black" but X is white, then we can do a coordinate transform such that X is now the name of a white object. Thus, every false statement can be converted into a true statement through a coordinate transform. However, if we do perform such arbitrary transforms then other statements which were previously true will become false. Now, we cannot judge which frame is the best one based on the truth condition, because there will indeed be some false statements in nature and they must remain false. Which statement is true or false, is an *outcome* of the language, not the method by which we can determine the language. In conclusion, there is no empirical method to decide the universal frame. We must take recourse into equating the naming with the meaning.

In Gödel's proof, Properties (O) → Name (O) where O is some given object or proposition. However, in semantic naming, Properties (O)

→ Name (O) → Meaning (O). In quantum theory, the properties of an object (mass, charge, etc.) lead to the wavefunction (in the position representation) which represents a new kind of *location*. Knowing this wavefunction is equivalent to knowing all the object's dynamical properties. Therefore, if the wavefunction was interpreted as a new kind of location, then this location defines the name by which we distinguish the object, then naming amounts to meaning.

Turing's Theorem

For a while, Gödel's incompleteness was an academic problem until Alan Turing showed that the problem of finding if a program will halt is undecidable. Turing's proof says that there is no program that takes another program as an input and decides whether or not it will halt. Herein lay the germ of another basic issue. Note that the definition of Turing's problem requires treating programs as both formulae and inputs to those formulae. Every computer stores data and programs as binary digits. However, these digits are interpreted differently, and data and programs are stored in separate memory locations. Turing's proof rests on passing a program as input to a program, which—he shows—leads to a contradiction.

To illustrate the problem with Turing's proof, let us assume that a program that solves the Halting Problem is called HALT and it accepts two types of inputs—P (the program) and I (input to the program P) to determine whether program P will halt on input I.

```
HALT (P) = 1, if P halts
HALT (P) = 0, if P loops
```

Assume also that we can formulate another program—TLAH—that does the reverse of HALT. If HALT says that P stops, then TLAH will loop forever. If HALT says P loops forever, then TLAH will stop.

```
TLAH (P) {
        if (HALT (P) { Loop } else { Stop }
}
```

A program is a number and can be passed as input to itself. This is like Gödel's case where a referring statement and its reference are represented by the same number (creating a self-reference). In Turing's proof, a program and its input are both represented by the same number. This is achieved by passing TLAH as input to itself.

```
TLAH (TLAH) {
    if (HALT (TLAH) { Loop } else { Stop }
}
```

Note that there are two possible instances of TLAH here—the program and the input. We might call these P_TLAH and I_TLAH. The following two cases are possible with respect to the outcome.

```
Case 1: If I_TLAH loops, then P_TLAH stops
Case 2: If I_TLAH stops, then P_TLAH loops
```

Since I_TLAH and P_TLAH are the same program, both alternatives imply a logical contradiction. Turing takes this to mean that the Halting Problem cannot be solved because the assumption that the problem can be solved leads to a logical contradiction.

Turing's proof invokes program semantics in a subtle manner. The question about whether a program halts pertains to whether the program constructs an object. This object can be given a meaning. If the program is meaningful, then it will construct a meaningful object, and it will halt. In current computing architectures, a compiler determines whether a program is syntactically correct but not whether it is semantically correct. Syntax compilation is quite like Gödel's numbering scheme that directly maps physical states to a number. A semantic compiler would first determine the meanings of each token before the meanings are mapped into a number. In effect, a semantic compiler will produce a semantic name rather than a syntactical name. While even meaningless ideas can be converted into numbers in current compilers, a semantic compiler will only convert meaningful programs into numbers. A semantically correct program must also be syntactically correct, but the reverse is not true. Semantic correctness is a stronger criterion and includes syntactical correctness. The halting problem therefore can be restated as the question of whether the program has a meaning.

When a program is passed as input to HALT, the program's meaning is passed. This meaning can be understood in one of two possible
ways. First, we can think of it as an algorithm; an algorithm is meaningful if it does not have infinite loops. Second, we can think of the
meaning as the problem that is solved by the program, or the purpose
it serves. These two kinds of meanings are implicit in Turing's proof as
the idea that a program represents functions and purposes. In current
computing, we cannot speak about either of these meanings; we cannot speak about algorithms and purposes because the computer does
not know the algorithm or the problem it is solving. And a computer
cannot therefore know if the program will eventually halt because we
can't know if the program has a meaning. If the program is meaningful
it will halt. So, the question of halting hinges upon knowing whether
the program has a meaning.

Turing's proof drops the distinction between a program and its
meaning without a semantic compilation procedure that determines
if the program is meaningful. Syntactical compilation determines syntax correctness, not semantic correctness. A semantic compiler that
determines semantic correctness will also ensure syntax correctness, although the distinction between a program and its meaning
will remain because knowing the output of a program does not tell us
how that output was constructed. As in Gödel's case, semantic compilation depends on a semantic view of space, except that we must
now recognize that there are two kinds of meanings—descriptive and
prescriptive. When we use the word 'car' to describe an object, we
intrinsically attribute behaviors to that object, namely that it can be
driven. If something is described as a 'car' but it cannot be driven, then
the description is false. This fact about the relation between concepts
and programs complicates the semantic considerations that need to
go into a physical theory. We must recognize that a symbol can denote
a concept as well as a program, and there are two ways to interpret the
meaning of a symbol.

In current computing architectures, programs and data are represented by bits. But the bits in the memory are interpreted differently
than the bits in the data memory. In current computing architectures,
this distinction is achieved by keeping data and programs in separate memory locations. The program (called text) cannot be modified

during program execution but the data can.

Since position and momentum are two *representations* of the same state, when the concept is modified its behavior must also be modified. In effect, this implies that if an object is transformed from a car to a shirt, then its accompanying program will also change from 'driving' to 'wearing.' If a program represents a concept, it will also be a meaningful program, as the two views are complementary.

Programs in physics appear as laws of nature. If physical properties—e.g., locations in space—denote the object's meaning, a particle's location can indicate the possible behaviors of the particle. These behaviors can be called the *abilities* in the object, which can be realized through a relation and under a purpose. Therefore, the behavioral description is not deterministic; it works only when there is a purpose to utilize the ability and the ability is exercised in relation to a suitable object. But within these constraints, it is still possible to treat quantum object as a program that *may be* executed. You can call it a *stored program* that exists but is not yet being executed.

Quantum Communication

If quantum objects are symbols, then quantum probabilities are the likelihood of finding a symbol in a text. An ensemble of quantum objects is an informational object like a book built of symbols. Probabilities of quanta represent the frequencies of finding different symbols. Books on different subjects, and indeed different books on the same subject will have different word frequencies. To that extent, quantum probabilities inform us about the measured system, but the knowledge is incomplete. Probabilities mean that a vocabulary can be used to construct many different statements. Probability refers to the amount of information that can be added to the quantum system after we have fixed the vocabulary. Quantum theory is incomplete because it describes the vocabulary but not the statements encoded using this vocabulary. The indeterminism during measurement is the new information received. This information was earlier encoded by the sender during state preparation. Quantum laws do not fix semantic novelty in communication and it is possible to transmit many meanings under the same laws. Viewed this way, quantum theory is the possibility of communication at the atomic level. This view of communication is also consistent with macroscopic intuitions about communication. Therefore, quantum theory can be the theory that explains communication outside classical physics.

Quantum ensembles satisfy some key everyday facts about information: (a) stationary states in quantum systems allow information to be encoded in a stable manner, allowing a receiver to measure exactly what was encoded, and, (b) discrete states corresponds to the use of alphabets in ordinary languages. Quantum theory, therefore, has more in common with everyday intuitions about information communication than classical physics. Classical physics violates the quantum idea of stationary states[1]. A classical ensemble with a finite amount of energy is never in a stationary state. If the classical ensemble denotes a meaning, then that meaning shall constantly evolve. The meaning that the sender encodes will be different from the meaning that the receiver will decode.

Probability is a problem if we think that quantum theory must predict the next position of a particle, because the next position seems the kind of thing that a scientific law must predict. However, probability is not a problem if the theory must predict the next letter in a sentence, because the content of communication is a choice by which the sender encodes a message in matter, which the receiver can decode. A time-dependent wavefunction will represent evolving knowledge— like a book that was changing its meaning. But a time-independent wavefunction will represent fixed knowledge as encoded by a sender for interpretation by a receiver. The apparent changes in particle states closely seem like motion, but they are not. Knowledge evolution involves change in information caused by the transfer of information (energy) from the source to a destination.

Communication requires time-independent stationary states where a sender can encode information which the receiver will decode without alteration during transmission. Evolution requires exchange of information between a system and its surroundings. The quantum ensemble in the first case is a closed system, which permits the creation of stationary states (i.e. fixed energy). The ensemble in the second case is an open system that allows some exchange of information, thereby changing the total energy in the system. Both closed and open ensembles are inadequately described in current quantum theory. Probabilities in a closed ensemble represent information encoded by the sender. Probabilities in an open ensemble pertain to how information is transformed dynamically.

The state preparation of the quantum object is like the act of authoring a book. The meaning in the book exists objectively, but the order in which you read the book is also important to determine the sequence of words. For instance, if you open the book on a random page, the order of the words will begin from that page. The quantum observations are however not random; they reflect the order previously encoded in the book. The encoding of the order itself may be not as precise as writing a book because we are dealing with macroscopic state preparations which don't necessarily set the microscopic states. However, whatever order was encoded previously would be observed during the measurements.

An important ingredient of causality here involves two new considerations. First, we must recognize that the encoded meaning exists as a *possibility*; therefore, it is not always manifest or observed. Second, we need to recognize that there is an agency that converts this possibility into reality; there are various proposed mechanisms by which this 'collapse' is supposed to occur, but I will argue that these are not the real mechanisms. The real causality is in fact not in matter but in *time*. The reason for this is that a quantum system settles into a stationary state provided it is not exchanging energy with other systems, and the subsequent exchange of energy itself involves the conversion of a possibility into a reality, which currently remains unpredictable and subject to the 'collapse' hypothesis.

This problem can be demystified if we give time a causal role; time will now collapse the wavefunction into a specific state or realize the possibility into a reality. For this to occur, time cannot simply be a linear dimension without causal effects. We must rather think of time semantically, which becomes possible when time is cyclic. For instance, in everyday notions of time, the day is divided into morning, afternoon, night, etc. and this passage of time itself causes a life cycle; we wake up in the morning, work during the day, and sleep at night. So, time has a causal property, when it is described as morning, afternoon, and night. Furthermore, these cycles in time are embedded into higher cycles, producing a hierarchy. Thus, the time of day is embedded in a month, year, century etc. which are longer cycles, and each of these cycles themselves have a meaning. For example, the seasons in the year have a meaning; they cause different weather, flora and fauna, and even change behaviors.

This hierarchy of time cycles produces nested effects. For instance, the effect of a moment in time right now is influenced by the properties of the day, month, year, century, etc. Each cycle is running at a different rate, so the embedding of cycles produces a combination of effects which will appear random if we are unfamiliar with the hierarchy of embedding. Furthermore, each such instant in time will trigger different possibilities into an observation. So, if the combination of effects of hierarchical cycles of time seems random, then the effects produced by these cycles will also be random. The surprising part is that if you look at things on average, you will always find a repeating pattern, which generates probabilities. These probabilities can be explained in two ways, only one of which is used currently. First, we can say that the individual events are truly random—i.e. we cannot explain why they occur—but they follow a probability. Second, we can say that the individual events have an order which is produced by a very complex process that we are unaware of—i.e. we can explain why they occur if we understand the logic that generates them. In current quantum theory, we treat the events randomly, but we can also treat them as outcomes of a complex process in time involving a hierarchy of causes.

If we begin the experiment now, the nature of the current time will determine what you observe. That is like opening the book on some random page. As time passes, we don't read the book linearly. We rather open a different page each time. We are still reading a pre-encoded book, but the process of reading is nuanced by the fact that we don't read the pages in a sequence. Time rather causes different pages in the book to be manifest at different times. Over time, we would have read the entire book and obtained the entire meaning, but the sequence in which we read would seem random. Thus, quantum experiments only superficially seem random due to the nature of time, and how it manifests possibilities into events. Factually, there is an underlying fixed reality—i.e. the book. And there is a formal method of specifying the order based on the hierarchical structure of time that manifests event sequences.

The scenario is somewhat further complicated by the fact that the quantum state preparations aren't very precise as they are done using macroscopic procedures. Therefore, the amount of meaning that has been encoded in a macroscopic system is far lesser than the amount of meaning that can be encoded. This gap between the total possible information and the encoded information creates an apparent randomness, because the gap corresponds to randomness. This randomness is further compounded by the seeming randomness in time. The former type of randomness cannot be overcome unless we can do precise state preparations, but the latter form can be overcome if we understand the hierarchical cycles in time.

This view of quantum measurements leads to a new picture of reality in which matter is the *message* between observers. Through state preparation an observer encodes a message in matter which another observer can read. The message that the receiver obtains is the message that the sender encoded. To encode messages using quantum objects, state preparations will also have to be more precise. Such object manipulation is impossible in current physics because atoms are treated as things rather than symbols. To program an ensemble with meaning requires a method that can encode meaning rather than just transfer energy. What current quantum theory describes as the superposition of states will in the new theory represent a combination of ideas into propositions. The new theory gives us the physical basis for encoding propositions wherein ideas are created by combining other primitive ideas. An ensemble of quantum particles is a system for encoding all propositions that can be created by combining atomic symbols within the ensemble.

Measurements and Information

Measuring instruments are classic examples of informational objects because a measuring device has both physical and symbolic states. At the point of a measurement, the measuring device registers pointer movements which indicate a change in the state of the measuring instrument but also in the state of the measured system. The pointer movement is symbolic because it is not simply a physical property of the measuring device, but it indicates the state of the measured objects. If we treated pointer movements as ends in themselves, we could plot the probabilities of a pointer moving by a certain value— let's say in a grocer's shop. That is where quantum theory stands currently, as it gives event probabilities without an explanation. But if we had to explain these probabilities, we would correlate pointer movements to different types of things being weighed on the scale and obtain new 'laws': spices weigh 10 gm, biscuits weigh 100 gm, flour weighs 1 kg, etc. We still have probabilities, but now we have obtained useful facts about the world we are describing: 1% of all groceries by weight are spices while 30% are flour. That description is more meaningful than simply the percentages of various pointer movements on the weighing scale.

Pointer movements are meaningless unless correlated with the corresponding events they measure. In present quantum theory we have event probabilities, but we don't know what we are measuring. This theory won't be completed by additional facts, because we already have all the facts obtainable from the measuring system. What we need now is the ability to correlate these measurements with the nature of the measured objects. Quantum theory will be completed when we are able to understand the nature of what we are measuring. This is possible if we can distinguish the things being measured using concepts and programs—quite like correlating weights given by pointer movement with the types of groceries.

What the above example illustrates is that the problems of quantum theory have everyday counterparts, and these aren't unique to sub-atomic particles. This insight is necessary to see that these problems will not be solved if we limit ourselves to studying sub-atomic particles, because the intuitions for solving them are easily obtained in the everyday world. The problems of quantum theory arise because we are in a grocer's shop measuring the probabilities of a shopper's bag having a certain weight without looking at what the shopper is buying. A grocer's accounting would be incomplete if he only noted the weights without the types of things he is selling or if he only noted the types of things he is selling without noting their weights. A complete account of a grocer's sales requires the grocer to account for both the types of things and the quantities sold.

Since weighing scales can't detect the type of thing being weighed, groceries are marked with barcodes which describe other facts such as type and price. The barcode stuck on the grocery is information that completes the accounting. Information in the barcode is not physical properties and yet it can be *conveyed* to us as physical properties. In other words, it is possible to interpret the barcode ink as a possessed property of an object, but it is also possible to see it as information that is not exhausted by physical properties. Someone measuring the barcode as a physical property won't be wrong, but he or she would miss out on viewing the barcode as a *symbol* and thus gaining knowledge from the measurement. Likewise, the eigenfunction of an atomic object could be viewed as a physical property. But treating these properties physically we would miss the fact that it can be a symbol representing meaning.

Substance versus Form

Plato believed that there is a world of *forms* above us and it consists of all pure ideas that we can think of. The pure world of forms includes, for instance, the idea of a pure chair, pure table, pure heat, pure light, pure mass, pure momentum, pure energy, etc. The present imperfect and impure world of things reflects this pure world of forms. This otherworldly view lies at the root of the philosophical problems that have raged on for the last over two millennia. The difference between the idea-like Platonic world and the thing-like material world became the mind-body problem during the time of Descartes and continues to reappear in theories of intelligence and the mind. The split between idea and matter is also at the root of the problems of quantum theory because now objects can convey information beyond their possessed material properties.

The Platonic views about the relation between matter and concept were revised after Plato. The key issue was the supposed other-worldliness of the ideas and lack of clarity around how these ideas are reflected in the present world. Aristotle, who succeeded Plato, revised his stance to include both concepts and matter within the same reality. In Aristotle's philosophy there are two things—substance and form. Substances are amorphous, and they are carved into objects by applying form to them. Forms are like Platonic ideas, but they now belong to the present world. Aristotle believed that every material object can be divided into two parts—the "stuff" of which it is made and the abstract idea that it embodies. Thus, for instance, a block of stone is carved into a statue by applying the *form* of the statue to the *substance* of stone. If the same statue is carved using plastic, then the substance will change but the form will not. In a swift move, all ideas from Plato's world descended into matter. To know these ideas, we only had to empirically observe objects.

The interesting thing about physical shapes is that they can denote both conceptual and physical properties, although conceptual properties always arise in a contextual manner. Aristotle exploited this dual interpretability of forms. Within a chair, for example, the shape denotes the physical properties of the chair. And yet, in some sense the shape of chair is like the abstract form of an ideal chair, and therefore a representation of the idea of an ideal chair. This allows a shape to be treated both physically and conceptually. The problem in Aristotle's view is that it separates substance from form, which means that we can no longer apply concepts to substances. Thus, for instance, we could say that the substance was 'clay' or 'plastic' because these would then become concepts—i.e. forms. So, how do we study these substances? If all material objects comprise these substances, then matter cannot be studied using concepts.

The next important issue about forms (which Plato and Aristotle did not consider) is whether forms are fundamental or emergent properties of nature. The question was not very important at that time because atomism had not taken hold of Greek thought (even though Democritus proposed atomic theory before Aristotle). This question becomes very important with atomism because if matter is divisible into smaller particles, then these particles are already individuals, like chairs and tables, although much smaller. Should we now suppose that these atomic objects themselves have a form that is fundamental? In current quantum theory, atomic objects don't represent types although we believe that when these particles are aggregated into macroscopic objects they come to represent types like chairs and tables. In current science, therefore, forms are emergent (and not fundamental) properties of ensembles and evolutionists would say that they emerge accidentally. Philosophers in fact even debate whether there is anything real called form, even at the macroscopic ensemble level. They might argue that fundamental particles don't have a form and the macroscopic world *appears* to have a form because we *interpret* the world that way. Thus, we perceive form and function in the world, but only the atoms are ultimately real. This leads to the idea that matter has no meaning and cannot represent meaning at fundamental levels of reality. However, collections of atomic objects can be meaningful in an everyday sense, but this meaning has no scientific relevance.

Quantum theory changes this because its incompleteness requires a new explanation of probabilities and it cannot involve physical properties. The dual interpretability of forms as physical shapes and as conceptual meanings can solve the problem without requiring additional properties. We would have to just get rid of substances! For instance, even substance—such as plastic or wood or clay—is also a form. The world is therefore entirely comprised of forms; we give these forms different meanings. Some of them are treated as substances and the others are treated as ideas. Forms are included in the current theories, so these are not additional physical properties. A reinterpretation of forms as meanings can yield new explanations in physics without exiting the empirical framework of science. In this reinterpretation, forms in matter denote not just physical substances but also conceptual universals. What we call 'matter' is therefore a collection of forms or simply concepts.

The notion that matter is meaning has implications for the mind-body problem. If forms are not in the other world as Plato claimed, then the problem of how mind understands meanings can be solved by supposing that the senses grasp the possessed properties of objects while the mind grasps the relations between objects. Meanings can be encoded in brains just as physical objects can carry meanings. Both the brain and physical objects are semantic in the same sense. This considerably simplifies the problem of how brains interact with the world, although it requires the postulate of informational interaction between the brain and the world. If the brain and the world interact informationally, then objects themselves can also interact through information exchange. This changes how we view the mind, but also how we conceive of material causality. Causes are now based on infor-mation exchange rather than force.

Meanings and Distinctions

Unlike physical properties that are defined universally—and each physical property can be perceived even when other properties are absent—meanings can be perceived only through a *distinction* between opposites. Red denotes risk through its opposition to green and yellow which denote safety and caution. Red denotes masculinity in opposition to pink or purple, which denote femininity. Meanings associated with red cannot, therefore, be known unless other oppos-ing colors are present in the context. In that sense, the meaning of red-ness is clarified within some context, through a relational distinction between different colors. While these distinctions are interpreted by the mind, they are not necessarily subjective. The meanings of colors, for instance, span across languages and cultures. Like current science objectifies physical properties based on sensations, meanings can be objectified in a semantic science.

Human conceptual development progresses from things to sensa-tions to concepts. When a child associates the word 'square' with a square object, he or she thinks that the word 'square' is the *name* of a specific object. At this time, only a specific object is called a square. As the mind develops, the child understands that square refers to the physical property of an object when the object has four equal sides at right angles to each other. At this time, only perfectly square objects are called squares, but several objects can be square. When the mind develops further, squares are understood by their distinction to circles, rectangles, rhombuses and other polygons. Now, the child will call even imperfect squares (such as when all their sides are not exactly equal, or sides may not be at right angles) squares. The perception of irregular shapes in terms of regular shapes requires a contextual distinction. As illustrated in Figure-3 the shape in the middle is a circle as compared to the shape on the left but a square as compared to the shape on the right. The fact is that from a perceptual standpoint, the shape in the middle is neither a square nor a circle. The properties of being a square or circle are additional properties associated with the shape relative to a context. The word 'square' thus denotes a universal and objective property of things that have equal sides at right angles. But the same word also denotes concepts that can contextually distinguish objects.

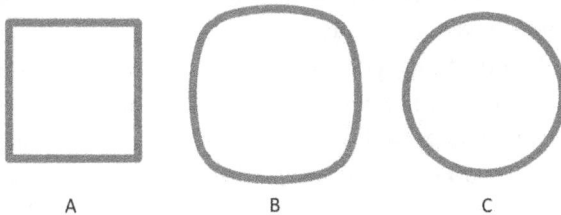

Figure-3 Distinctions and Concepts

Another classic example of such semantic properties is seen in the cognition of frequencies as musical notes. A frequency by itself is not a note, unless the frequency is observed in relation to other notes. A collection of notes forms a scale, after which frequencies are also labeled by signs as Do, Re, Mi, Fa, Sol, La and Ti. While musicians often associate standard frequencies with notes to synchronize different instruments, musical perception does not depend on the absolute frequency scale but on the ability to see distinctions between frequencies. While most observers can recognize higher and lower tones, they cannot always perceive musical notes in them. The perception of musical notes in frequencies is a semantic perception. Musical distinctions are universal, even though not everyone is musically conversant. To see notes in frequencies requires an additional layer of meaning above that of physical properties.

Semantic Space

Perhaps the most profound example of semantics exists in the ability to use the physical property of location to denote meanings. In

Figure-4, locations on the X axis represent colors based on increasing frequencies. Each location acquires an additional meaning when the axis cutting through the points is interpreted as frequency of visible radiation. This fact is used in physics when spatial locations are used to distinguish objects based on properties other than their position. Points in space can represent distinctions between mass, charge, momentum, energy, etc. Sometimes points in space are color; at other times the point can represent taste, form, smell, or other properties. Cartographers draw maps of the world on paper, where locations on the space denote cities, countries and continents. Engineers draw designs of machines on paper and architects draw construction plans. Computer scientists draw flow-charts and object relationships using space. These and other uses of space associate locations in space with different meanings. Pixels or points in space can denote a variety of distinct properties depending on the context. In these drawings, the physical property is just a location in space. But locations are additionally associated with contextual meanings.

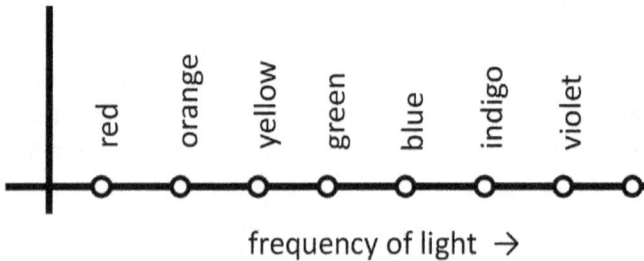

Figure-4 Spatial Locations Convey Information

Every concept can be represented through distinctions. And every distinction can be expressed via a spatial *distance* between points[1]. Every distinct object must have a distinct location (at that point in time) if it is a distinct object. Two objects that are located at the same point in space are, essentially, the same object. Conceptual distinction between objects can therefore be represented by a distinct location. This fact has been used in the last three centuries to represent a variety of scientific information using conceptual spaces. Points in these conceptual spaces represent individual objects and their properties. Such conceptual spaces are often drawn on paper, whiteboards or magnetic media. Conceptual distinctions on paper, whiteboards or other kinds of media are overlaid on spatial locations in everyday 3-dimensional space. So far, science has given a variety of meanings to points on paper or whiteboards, while not giving any meaning to the points in the underlying space. By this bias, we suppose that paper can represent information about particle trajectories, but material particles can't represent meaning.

What is the difference between a paper and a classical particle? Can we not treat a paper with trajectories drawn on it as a classical particle? Classical physics forbids this. When we reduce a paper to a particle, the former loses all internal content. The classical particle can only tell us about itself, and not about anything else. By reducing the piece of paper to a particle, the drawing on the paper will be reduced to measurable properties of the paper itself. Measurements on the paper drawing will convey properties about the paper but not information about other particles. Note how information conveyed by a paper drawing does not require a new reality apart from what is already obtainable via the measurements themselves. Treating the paper drawing as information merely requires another way of knowing the world, where distinctions in space are not interpreted as physical properties but as knowledge about other things.

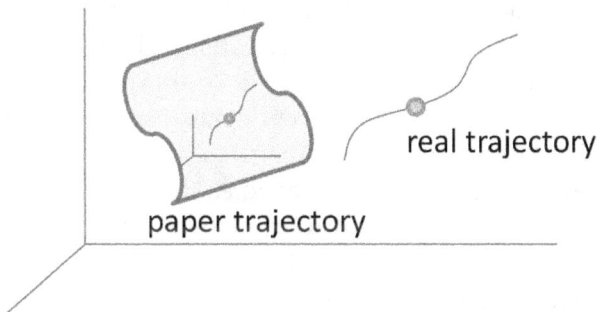

real trajectory

paper trajectory

Figure-5 Pictures vs. Classical Particles

Since all distinctions are represented via spatial distances, information merely requires a reinterpretation of space. When an axis is
interpreted as the distinction in color, then points on the axes denote
various colors. When the same axis is interpreted as the distinction in
shape, then points on the axis denote various shapes. The points and
their mutual distance (the physical properties of the points) remain
unchanged through such interpretations. But, by interpreting the
space differently, science can give the same objects newer meanings.
Specifically, objects can now convey information about other objects.
Instead of being a line on paper, the line now starts to represent a
trajectory when the space on the paper is interpreted as the 3-dimen-
sional space in which an object moves.

This leads us to the conclusion that ordinary space is also semantic
under a different way of looking at space. This fact has been used for
centuries when drawings on paper were interpreted as particle trajectories. What is now needed is to stop thinking of the piece of paper
as a classical object! The paper that conveys a picture is as much a
quantum object as electrons. Classical physics removes the information that a piece of paper can provide, and it thus gives us an imperfect idealization of reality. Every object that conveys information is a
quantum object because it provides knowledge. Quantum phenomena
are not obvious when the knowledge provided by an object is identical to what we can derive from its sensations. They become obvious
when knowledge and sensations are different. But every object provides both knowledge and sensation.

The quantum 2-slit experiment[1] can be seen as the physical realization of the possibility of associating locations with meanings. Quanta
arrive at distinct points in space in the 2-slit experiment not due to
randomness. They arrive at different points because quanta denote
meanings. The distance between spatial locations can denote a color
axis in one case and a shape axis in another. The same location in space
can sometimes represent the distinction between color and at other
times the distinction of shape. This meaning of locations is beyond
what current mathematics or physics sees in points. To incorporate
additional views of space-time in physics, therefore, quantum theory
must go beyond classical notions about position.

Quantum indeterminism can be explained if the positions of

quantum particles are taken to represent concepts. These concepts may use the same words as sensations, but the meanings are defined relationally. In representing colors like red and yellow, quantum objects will not provide the sensations of red and yellow. They can however express the distinction between yellow and red and hence *knowledge* of yellow and red. The quantum experiment is like reading a book which has physical properties, but those properties are also tied to meanings. Put simply, a quantum experiment gives knowledge different from the sensation. To explain quantum indeterminism physics needs to use the knowledge in addition to the sensations. The interesting fact about concepts is that concepts of touch can be expressed through sensations of color, the concepts of color can be expressed using sensations of taste, the concepts of taste can be expressed using sensations smell, etc. In these expressions, there is a difference between what is being sensed and what is being known. The predictions of the scientific experiments must be based both on the semantic conceptual content and on physical properties.

The meaning in that object is not some material stuff that we added to the object during state preparation. The meaning is simply its location in space, when space has been given a new kind of meaning. Since every object is in space, all objects can convey information when space is re-interpreted, and all objects can be quantum objects. The quantum behavior is not limited to atomic objects. In fact, the mystery of quantum theory is not even in matter, whose atomicity we believe to be studying. The quantum mystery comes from the ability to interpret space as various kinds of property axes and locations in space as various kinds of distinctions. Through such interpretations, two objects on that axis can convey information beyond their physical properties. A symbol may use an object's physical properties to represent information, but those properties are simply the tokens and not the meanings. The knowledge we get from symbols is more than the tokens. The indeterminism in current theory is due to the gap between knowledge and sensations—we are describing quantum objects in terms of the sensations they offer while neglecting the knowledge represented by them.

To solve this problem, we must treat space semantically. The locations and directions represent meanings. A location can represent color or mass, which converts an object into a symbol. On a paper drawing, we label an axis by a property name ("mass", "energy", "color," etc.). This labeling interprets space, and points on paper become symbols. By moving the particles, we will change the meaning encoded in the ensemble. When an ensemble is observed, the meaning derived from measurement depends on how locations of objects are labeled. For instance, in the 2-slit experiment, the distance between detectors could be used to represent differences of color, taste, smell, form, and any other kind of measurable property. The quantum detections can be explained if we can postulate the ability to convert length into the dimension of some property.

Higher-Order Logic

Intuitions in physics have thus far been drawn from the study of object interactions. These created the notions of particle and wave, and they inadequately explain quantum phenomena. A new set of intuitions can now be sought from a different type of objects—namely books, pictures, music and science, rather than particles and waves. We believe that minds *perceive* meanings in books, music, pictures and science rather than manufacture it. That is, meaning is in the books, pictures, music and science and it was embedded by their creators. To embed meanings, there must be at least one natural language that can be mutated by different creators. Communications between material objects must rely on this language.

Physicists already believe that mathematics is the language of nature, but this language uses first-order logic—the description of quantity measurements. In first-order logic, we cannot hypothesize the existence of *objects* that have physical properties, because that requires second-order logic which distinguishes between objects and their properties. Notions of realism in science are thus misplaced because they assume objects and their properties when first-order logic only deals with properties. This has however not created a serious problem in science because the notion of an object (apart from its properties) does not explicitly enter science. Objects remain metaphysical assumptions in science. They have some pragmatic usefulness in so far as they help the scientists visualize the world. Objects however become *necessary* in semantic theories because of the distinction between a symbol and its meaning. Now, we have two things—the physical identity of an object and the meanings of that object. We might now postulate that the physical identity *is* the object, but the meanings (properties) of the object are interpretations of the object. The object-property distinction now becomes the object-meaning distinction. While first-order logic sciences could live with only properties, semantic theories necessitate both properties and objects forcing us into second-order logic. However, now, the properties of an object are its meanings.

One corollary of Gödel's Incompleteness theorem is that no second-order logical theory is complete and provable. That is, if we introduce meanings into language as attributes of symbols, then statements made using the second-order logic will not be provable. This also means that there cannot be a formal theory that contains both objects and their properties. Current science only uses properties, which scientists attribute to objects, informally. But this idea cannot be consistent because objects and properties would allow us to have meanings as the properties of the objects.

Fundamentally, science needs to solve the problems of realism where both objects and properties are real because only then can we speak about symbols that have physical properties that are interpreted as meanings. It follows that a semantic view of quantum theory requires a new type of mathematics where we can speak both about quantum objects and their dynamical properties. In current quantum theory, we can speak about an object's dynamical properties but there is no conception of the object itself. Without a proper notion of a quantum object, problematic features of quantum theory such as uncertainty, probability and non-locality will remain problematic. These features can, however, be resolved in a semantic view where the dynamical properties are attributes of symbols. But that viewpoint itself requires both objects and their properties, necessitating the use of second-order rather than first-order logic.

If meanings are real, then matter should be described by second and higher order logical languages. Current physics is a first-order logical science; it describes physical states only. Using this approach, physics cannot describe objects with meanings. A semantic object has physical properties, which also denote its semantic properties. The distinction between and object and its properties necessitate the use of second-order logic which is presently impossible unless we induct linguistic categories like meaning and naming into mathematics. Numbers are currently representations of naming. Second order logic therefore requires the induction of meaning as well.

The intuitive basis of understanding quantum theory, I believe, lies in understanding mind-like phenomena in matter, without bringing the conscious observer into science. Quantum theory does not pertain to the ability of sensation or thought in observers. It rather pertains to concepts and algorithms, which are objective but quite different than objectivity in current physics. Whereas objects so far had properties independent of other objects, a new kind of objectivity is now needed that conceives of objects within a contextual ensemble of other objects. The contextual properties are objective in the sense that they exist in matter whether we observe them or not. However, their existence is subject to the existence of other objects in the ensemble. In that respect, objectivity is alive in quantum theory although classical reductionism that treated an ensemble as a combination of independent particles has failed.

Hierarchical Addressing

The intuitions discussed so far lead us to the conclusion that locations in space become symbols of meaning when distances between them are interpreted as a property axis. The directions also become representations of types (e.g. opposite directions can be opposite types) when angles between them indicate similarity of type against the chosen property axis. Thus, we can define types in two distinct ways—as different properties and as different values. The distance can represent the type and the angle can denote the value.

The shape or form of the eigenfunction is now a physical property, which denotes the form of a symbol. However, the dynamical properties of the eigenfunction represent meanings. To denote a symbol, we must allow extended objects. As Descartes said, all matter is *res extensa* which means that because matter exists in space, it is always extended, with a position. Classical physics reduced matter to point particles or infinitely extended fields. Points don't have a shape or size because they are infinitesimal, and fields can't have shape and size because they extend over the entire universe. In quantum theory, forms, locations and directions are contextual. This means properties of space can be treated semantically in quantum theory. The shapes of eigenfunctions in quantum theory can represent symbols, while the dynamical properties of the eigenfunctions will denote meanings. We presently treat these shapes as an uncertainty in a quantum object's location. We could also treat the shape as a symbol. The type of the object can be given by extension (shape), which is a physical property that *conveys* information content within the context of an ensemble.

I am suggesting that nature acquires a symbolic language in a context due to which dynamical properties of objects indicate meanings unique to that context. The context—in this case the ensemble—defines the language of signs in which meanings are mapped to dynamical properties. The ensemble is a collection of objects, but inside the ensemble, individual particles represent logically orthogonal meanings—like alphabets in a language—such that the meaning of the sum of alphabets represents a meaning in the proposition. The supposed superposition of eigenfunctions in the wavefunction represents, in fact, a combination of alphabets.

The eigenfunction has observables but the eigenfunction is not those observables. In SI, the observables of quantum theory are different types of meanings. The eigenfunction is the symbol of these meanings. That is, the same symbol has many kinds of meanings; for instance, a symbol can denote a description and a program. The reality from which these meanings are derived is the eigenfunction and we can postulate that the eigenfunction exists as objective information which gives rise to dynamical properties when observed. The dynamical properties can be interpreted as meanings if space and time are viewed as domains of meaning. These meanings will include both sensations and concepts. That is, we will learn from the meanings not just that there is a stop sign but also that it is red. This will not just solve the problem that current physical theories don't describe concepts, but also the problem that they do not describe the world in terms of the perceived secondary properties.

The classical physical interpretation of the observable is not equivalent to the sensation of the property because the classical interpretation describes the world in terms of position and momentum while the senses observe it in terms of color, taste and smell. The classical interpretation of observables defines these properties in a non-contextual and universal manner. Since quantum observables are contextual, they should be interpreted as meanings. However, if we interpret them purely as concepts (such as a stop sign) then we have lost the ability to derive sensations from them (e.g., redness). The observables therefore must be treated as giving us both sensations and concepts. I will later describe how quantum observables could be separated into sensations and concepts.

The basic insight necessary for this distinction is the fact that our senses can manipulate much larger objects as compared to our minds which can think or imagine much smaller things. The meaning in a concept is therefore an individual quantum object, but classical physics only manipulates collections of such quanta. The dynamical properties of a particle *ensemble* can represent sensations although the senses cannot perceive the smallest quantum itself. Current quantum theory describes ensembles because we cannot manipulate the individual quantum. However, if the ensemble and the individual quantum represent information then it is not difficult to see that the individual quantum could be a representation of a concept or a sensation, although these would be microscopic in nature. Inevitably this will imply that when atomic *concepts* are combined macroscopic *sensations* are produced. The macroscopic and microscopic objects can thus both be viewed as information. And they can both signify concepts— some of which are sensations and others are non-sensible. Thus, the sensation concept can be redness, while the non-sensible concepts can be passion, war, love, life, danger, etc.

If this idea is correct, then the classic divide between sensation and concepts is incorrect. Both sensations and concepts are ideas; however, sensations are more 'gross' while concepts are 'subtle.' This difference between gross and subtle can be understood if we organize these ideas into a hierarchy—i.e. an inverted tree with the root representing the most abstract ideas while the leaves expressing the most contingent ideas. Our senses too must be treated as ideas. These senses can measure what is more contingent than the sense itself. The mind too must now be treated as an idea, although more abstract than the senses. Thus, the mind can grasp sensible ideas such as redness, just as the senses. However, the mind can also grasp subtle ideas such as apple, chair, and table, which the senses cannot. Thus, the ideas aren't independent of each other. The tree structure organizes them into a hierarchy and makes them perceivable in different ways. Each sense can perceive what is *lower* in the hierarchy than itself (i.e. farther away from the root). This allows the mind to perceive both sensible and non-sensible ideas, but the senses can only perceive what is sensible. Both can be viewed as ideas.

When we divide matter into smaller parts, there comes a point where the senses cannot *perceive* the objects anymore, and the mind cannot give them any meaning. Notions of quanta such as electrons, protons and quarks are *concepts* which are beyond the sensual and mental capacities. In fact, they are further beyond the mental capacity than they are farther from the sensual capacity. Thus, for instance, the effects of atomic objects can still be sensed when these quanta interact with macroscopic ensembles. However, even though we can sense these particles, we are still unable to understand them. The quantum interpretive problems arise due to this fact.

Therefore, it is important to go back into the everyday world and obtain intuitions about objectivity and meaning, and then extend it to atomic objects. The dynamical variables in quantum theory are derived from the *symmetry* properties of space and time. These have a counterpart in classical physics because in both theories the properties are derived from the symmetry properties of space and time. Therefore, the dynamical variables don't need to change. However, their classical interpretation needs to change because quantum dynamical variables don't commute, are complementary descriptions of the same state, and represent stationary states[2] rather than motion. The reinterpretation of quantum dynamical variables is possible if space and time are viewed semantically rather than physically. The dynamical variables can represent meanings rather than states of motion as they did in classical physics.

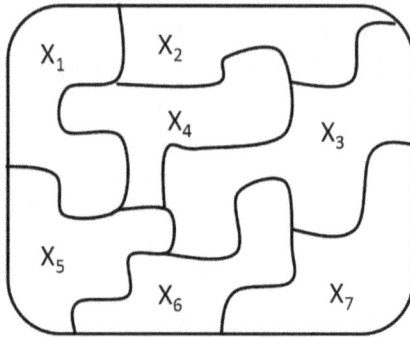

Figure-6 Space Quantization

However, we still need to maintain the idea that an ensemble of particles also has dynamical variables—e.g., position—because if we get rid of the ensemble's dynamical variables then we will lose the ability to explain why particles 'inside' the ensemble are entangled. We need a whole-part relationship between the ensemble and the quantum particles. The quantum particles are 'inside' the ensemble, but the ensemble is also a 'particle.' These particles are further organized in a hierarchy, so the dynamical variables of a higher particle denote the meaning of the larger subsets of the macroscopic objects as opposed to the same variables of a lower particle.

I shall call such forms in space *spacelets*, because of applying quantum theory to space. From a macroscopic perspective, the spacelet represents an ensemble of point locations. From the quantum perspective (in which the individual object cannot be divided into further smaller parts) the spacelet denotes an indivisible form. Figure-6 illustrates an example of an ensemble of particles, which could themselves be ensembles. Classical physics counted objects in a flat space-time and there could not be a particle inside a particle. In quantum theory, however, there are objects inside objects. The coordinate system for counting particles in quantum theory cannot be a flat space-time like in classical physics. Rather, objects must be counted hierarchically in relation to the ensemble that contains them. A spacelet *X1* is thus a *unique* location as well as a *collection* of locations. *X1* is labeled in relation to the containing ensemble and parts of *X1* would be labeled in relation to *X1*, not in relation to the container. This means that we must label objects in a *hierarchy*, quite like the labeling in the everyday world.

Think of how an apartment postal address is defined. Each apartment has a number, defined in relation to the apartment complex. The apartment complex has a number in relation to the street. The street has a name in relation to an area, which has a name in relation to the city, which is named within a state, which is named within a country, etc. We don't directly count the smallest particles in the universe. Instead, we count the objects through a hierarchy. This is important because in a flat space all objects would be counted independently. Only within an ensemble can the parts be given semantic meanings. Space thus cannot be flat and open in quantum theory if quantum theory is to be a theory of meaning. Rather, we must now construe a hierarchical theory of space, in which objects are numbered like postal addresses in ordinary use.

Time, similarly, must be defined in a hierarchical manner. Everyday notions time, such as 2014/08/16, 14:22:35 are hierarchical. Note how 2014 is present for the entire duration of the year, even as the infinitesimal moments pass. This notion about time can be used to understand how a quantum particle seems to go into the future and the past, even though it exists right now. A hierarchical notion about time can be used to define a *timelet*, which represents an extended duration of time within longer durations.

A spacelet is a region of space that sits somewhere in a hierarchy and can be given a label in relation to other higher spacelets. Spacelets form a class hierarchy and they are numbered using that hierarchy rather than independent of the hierarchy. A timelet is similarly a duration of time that sits somewhere in the hierarchy of temporal durations and can be given a label in relation or distinction to the higher level timelets. Unlike current quantum theory that describes matter and energy *in* space-time, a new quantum theory would describe matter and energy as the quantization *of* space-time. A symmetric space-time cannot be observed. However, if this space-time is quantized, asymmetries are added to space-time as discrete locations and times.

These asymmetries in space-time can be called *information*. They would represent sensations and meanings and analyzing them in a new way can create a new theory of nature.

4

The Semantic Interpretation

The lesson to be learned from what I have told of the
origin of quantum mechanics is that probable refinements
of mathematical methods will not suffice to produce a
satisfactory theory, but that somewhere in our doctrine is
hidden a concept, unjustified by experience, which we must
eliminate to open up the road.

—*Max Born*

Current quantum theory describes an ensemble of quantum objects, and not individual objects. This view is called the Ensemble Interpretation (EI) of Quantum Theory and was originally proposed by Einstein. Einstein's primary objection to quantum theory was that the theory does not describe the individual particle, but only a collection of such particles. The wavefunction which is a description of the ensemble does not therefore describe an individual object. The wavefunction is useful for making correct predictions, but the theory would be superseded in the future by something that provides a description of individual objects. This view is often misconstrued as a classical tilt in Einstein's thinking, especially after Bell's Theorem, which showed that hidden variables cannot improve the quantum theory. It is now well recognized that quantum theory is empirically complete. I will however argue that the theory is still causally incomplete because it does not explain all the empirical facts. Quantum events are currently explained probabilistically, but the order of the events is not explained, although the order is empirical. There can be a different theory of quantum phenomena that bridges this incompleteness, without classical concepts. Such a theory will also be free of the myriad problems that classical concepts bring.

EI isn't very popular because alternate conceptions about the individual quantum object have not been created. In lieu of an alternate view about quanta, present interpretations treat the wavefunction itself as the description of the individual quanta which leads to the idea that if a single quantum can have many states then it must be in a superposed state prior to measurement. SI provides that alternative conception of quantum reality. In SI, the individual quantum objects are information-carrying *symbols*. The order in quantum events is like the order between words in a sentence. It cannot be predicted without knowing the sentence in question. If we prepare quantum ensembles without attention to the individual quantum, the behavior will remain probabilistic. This observed behavior, however, hides in plain sight the possibility that we could prepare the quantum system like we construct a sentence, and it would then behave deterministically rather than statistically.

To understand this better, let us compare atomic objects with letters in the English language. English letters don't have meanings, but we know how to classify these letters into *types*—such as the ability to know that some letter is 'A.' The type of the letter is an analogue of a concept. Probabilities of finding a letter in a book can now be compared to probabilities of various measurement outcomes in quantum theory. If we describe a book in terms of frequencies of alphabets, we can assign frequencies to each of the letters as shown in Figure-7. But analyzing a book into letter frequencies does not tell us what the book means. Such frequencies may be invariant across books on fiction, poetry, technology, sciences and philosophy, although the meanings in books on different subjects are different.

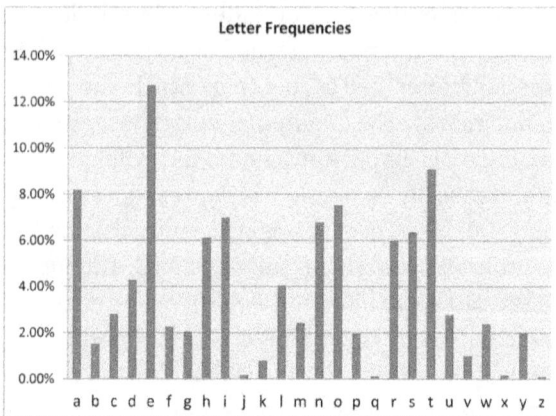

Figure-7 Letter Probabilities in the English Language

This analogy is not perfect because English letters aren't meaningful. A better analogy could involve words, but because words are much more numerous I chose alphabets to depict the idea. Both word and letter analogies are identical in that symbol probabilities do not convey the complete meaning. Part of the meaning is also expressed through symbol sequencing. In the case of written and spoken language, the forms of words and sounds are interpreted as meanings. These shapes and sounds are symbolic because they stand for something besides the symbol. The meaning of a symbol pertains to something outside the thing-in-itself; and yet, the meaning is based on the thing-in-itself. For instance, the meaning of a word indicates a concept or thing different from the word. And yet, the meaning is associated with the shape or form of the words themselves.

In quantum theory, this distinction between word and meaning can be compared to the distinction between the eigenfunction and its observables—like position, momentum, etc. Current quantum theory however interprets these observables classically. For example, the momentum of an eigenfunction is treated just like the momentum of a classical particle. The problem is that we cannot consistently treat the quantum observables classically due to uncertainty, non-locality, probability, etc. So, we are stuck. We can neither make quantum theory consistently look like classical mechanics, and we are unable to give these observables a new kind of interpretation.

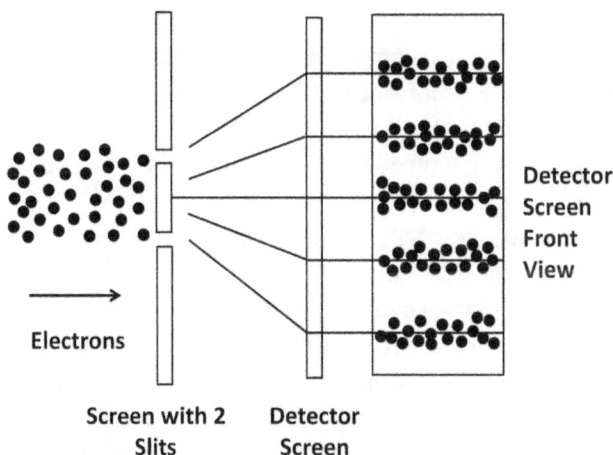

Figure-8 The Quantum Double-Slit Experiment

In the quantum double-slit experiment (shown in Figure-8), for example, quantum objects arrive at different locations on the detector. Quantum theory predicts the probability of finding an electron at a position, but all these electrons have identical possessed properties (e.g., mass and charge). No physical measurements of the electron's possessed properties can explain why two such identical electrons arrive at different positions on an array of detectors.

We could explain the different positions if we treated the electrons arriving at different locations as different *types* of particles; the difference between them being given by a different eigenfunction state. Since quantum theory predicts the probabilities based on the eigenfunction state, rather than on the electron's properties, we can clearly see that it is describing the outcome based on a *state*. How a given electron got into that state is not in the theory's purview.

The different positions of electrons in the 2-slit experiment can be likened to the different alphabets in language. Not knowing which detector is going to fire next is like not knowing which letter is coming next in a sentence in a book. Our ignorance does not mean that the sequence of letters isn't definite. Rather, our ignorance pertains to the fact that we don't *know* the encoded meaning.

The 2-slit experiment defines a lexicon like vocabularies in language. The ability to express concepts through properties is well known in everyday language. But what causes the property to signify a meaning isn't well understood. For a set of physical forms to express a set of concepts, the differences between these forms must be *isomorphic* to the differences between concepts. For instance, if a perfect square and triangle represent red and blue, then imperfect squares and circles could denote red and blue in some other context, provided imperfections are introduced in a consistent fashion (e.g., by rounding off the vertices of the geometrical shapes).

It follows that the ability in the forms to represent meanings depends not on the individual shapes of words, but upon a collective interrelation between the words. To represent a set of words, these forms must be defined collectively in an ensemble rather than individually. The probabilities of individual detections represent the likelihood of finding a word within a lexicon. The exact order of word detection depends on the state preparation by which a symbol sequence was encoded that can later be decoded. The experimental setup that encodes and decodes meanings is like the act of communicating meanings through writing and reading of a book.

Note how the meaning in the symbolic expressions does not universally reduce meanings to the forms of the symbols. Symbols, rather, represent certain meanings within a specific context. That does not mean that the relation between a symbol and its meaning is arbitrary. We know that the concept of danger can be indicated by many different shades of red, but not by the color white. To that extent, the relation between a sign and its meaning is flexible but not arbitrary. The signs by which meanings are expressed limit the possible shapes of the symbols, but don't necessarily fix it. This property of signs is remarkable because it allows a sign to convey the idea of an apple without itself being an apple. However, the sign must have some salient properties of the represented meaning for it to even convey the idea. We might say that a sign contextually embodies the information about the object it is designed to represent, not a universal and context-independent signification of the meaning.

If quantum theory is about symbols, then the fact that quantum eigenfunctions are defined mutually and collectively pertains to the representation scheme. The quantum eigenfunctions represent a set of symbols whose meanings are fixed by the context. Quantum state preparation is a contextual encoding of information and the intended recipients of this message are expected to be familiar with the language using which the message needs to be interpreted. Without the context, there are several possible interpretations of each symbol. Awareness of context limits possible interpretations.

The arrival of different electrons at different positions represents not just a physical property, but a semantic difference between electrons as well. To understand that semantic difference we need to view spatial distances as semantic distinctions where location in space is used to represent a type of concept. The differences in the meanings of objects are now understood by seeing spatial locations as a measure of meaning. This meaning is empirical in the sense that it can be detected because the electron arrives at a distinct location, although to understand this as a separate property we need to adopt a semantic view of quantum particles. We can't predict this meaning solely based on an electron's mass or charge.

Semantic properties are defined by the mutual distinctions between electrons (in this case expressed via the distance between detectors). It does not matter if we slide the entire letter distribution to the left or the right, or expand or contract the spacing between the letters, if the relative order of the symbols is being maintained. There is no classically relevant physical significance to be attached to the locations. This is easily confirmed by taking the battery of detectors closer or farther from the double-slit. By changing the distance between the slits and the detectors, the position of the detector at which an electron will arrive will change. The relative *pattern* and *probabilities* of the arrival, however, will be unchanged.

The locations of quantum detectors in the two-slit experiment can be compared to locations within a house. Space within a house can be distinguished with words such as kitchen, bedroom, toilet, living room, study, lobby, etc. Once space locations have been distinguished, they can be ordered by a person who walks to all these locations within a house. The two-slit experiment locations are like the locations in a house, and their order is like the sequence in which one might visit the different parts of the house. The position of a quantum object isn't a position in the classical sense because absolute magnitude is important for classical positions but irrelevant to quantum measurements, where only the *relation* between the positions (rather than their absolute values) is important. This is like how the meanings of locations like kitchen or bedroom in everyday language aren't exhausted by an absolute size or place in space (although kitchens and bedrooms do have a specific size in each instance). The different locations associated with quanta are analogous to kitchens or bedrooms in different individual houses. Each kitchen or bedroom is a sign of some universal concept.

We need to recognize that space can be given a non-classical interpretation and it is used in non-classical ways in everyday life because we can speak of going to the kitchen or the bedroom contextually without defining their exact coordinates in Euclidean space. Conceptual divisions of space abound in our world, where space is divided by countries, states, cities, streets and addresses. We can speak of going home or to work, or of going to a different city and country and this is perfectly *meaningful* in everyday conversation.

Conceptual ways of speaking about space *complement* the physical ways of speaking. Words such as 'home', 'office', 'bedroom', have the same semantic meaning although the places may be different in each context. In the everyday world, the confines of our home and life define a closed system within which words are used unambiguously. Likewise, quantum theory represents a closed system in which objects can be denoted by conceptual labels such that a label points to a specific location in that context. Quantum theory is telling us something profound, namely, that there is a different way to look at space than we have used so far in physics.

Understanding the Double-Slit

The difference between classical and quantum experiments is in what we consider a measuring instrument. Thus far, we have regarded the battery of detectors as the measuring instrument, and this leads to the idea that a quantum object has a position, but it can't be predicted. I will instead argue that it is possible to view the two slits as a measuring instrument and the detectors are *representations* of the value of measurement. This shift requires a close examination of the idea of measurement itself. Measurements are required to distinguish objects. Physics measures objects to distinguish them against a property: the measurement of mass distinguishes objects against the property of mass. But another kind of measurement is needed for distinguishing among different *types* of objects.

The difference between quantity and type measurements is that quantity measurements are universal while type measurements are contextual. To measure the color of an object, for instance, a universal standard color chart can be defined. But the color can also have a context sensitive interpretation: it can for instance be interpreted as risk, war, passion, masculinity or life in different contexts. In these contexts, the meaning of red is what it represents rather than what it is. When red denotes risk, green denotes safety and yellow denotes caution. When red denotes passion, blue denotes indifference. When red denotes war, white denotes peace. When red denotes life, black denotes death. And when red denotes masculinity, pink or purple denote femininity. The color of red can be defined universally against a standard shade chart but the meaning of the color is not merely an issue of observer dependence but also of the context in which colors are interpreted. This is because the same observer can view the meaning of red as danger, life, passion, etc. in different contexts. While a single object suffices to define the shade of color, at least two (or more) objects are needed to contextually define meanings.

This insight about meanings can be applied to the quantum 2-slit experiment, interpreting the two slits as a binary distinction. Every meaning measurement requires at least a binary distinction and quantum phenomena therefore begin with two slits. Quantum phenomena of course are not limited to two slits and can be done with 3, 4, 5 or more slits. As the number of slits increases, the same ensemble is represented in terms of ternary, quaternary, pentanary or higher distinctions. This is like how numbers can be represented in binary, octal, decimal, hexadecimal and other representations in digital computers. The two-slit experiment is a binary scheme. A higher number of slits represents higher representation schemes. The difference between quantum and digital representation schemes, however, is that the quantum scheme creates a distinction of *types* while the digital representation scheme creates distinction of *values*. Interference with multiple slits is depicted in Figure-9.

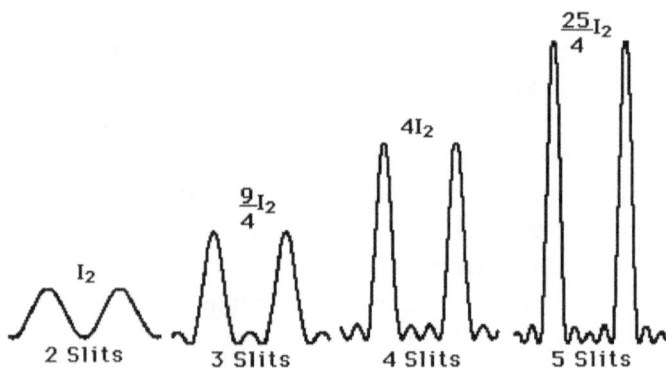

Figure-9 Multiple Slit Interference

The main mystery in the 2-slit experiment is why with two slits quanta never arrive at some positions which are allowed with a single slit. This was explained in classical physics by supposing that light is a wave and it passes through both slits simultaneously and interferes with itself. This hypothesis was proved wrong on two counts with the advent of quantum theory. First, it is not just light but also matter particles—e.g., electrons—that exhibit the same behavior. Second, a photon only passes through one slit and not through two slits as classical field theory postulated. The quantum is therefore always a particle and the observed 'interference' pattern must be explained with a particle theory and not with a wave theory. Since interference cannot be explained with the classical particle theory, a new type of particle must be postulated to account for the observed interference pattern in the 2-slit experiment.

In SI, quanta are symbols and their distinctness is one of *types*. The single-slit experiment measures type distinctions with just one type, which is clearly impossible because all type distinctions involve at least a binary opposite. In the case of quantity representations, the unary representation scheme will require infinite digits for each quantity, because no matter how many times we divide a number by 1, the remainder is always that number. The presence of infinite digits for a quantity leads to an infinite spread for each object, and we treat this spread as the wave-nature of the quantum. The problem, however, is in the measurement procedure, which tries to represent a finite quantity in terms of a unary representation scheme.

Like classical values can be represented using various representation schemes (binary, octal, decimal, hexadecimal), type distinctions can also be represented using binary, ternary, quaternary and higher distinctions. The single-slit experiment corresponds to a single *type*, like how classical physics treated all objects as *particles* (a particle is a single object type). In the 2-slit case, types will be described in terms of binary distinctions, e.g., hot-cold, black-white, etc. In 3-slits, types will be described in terms of ternary distinctions, e.g. red-blue-green, etc. The total amount of information in the quantum ensemble is identical regardless of how many slits are used. The number of slits, however, alters the representations of this information on the battery of detectors.

1001	$1*2^3 + 0*2^2 + 0*2^1 + 1*2^0$	9	Binary
1001	$1*3^3 + 0*3^2 + 0*3^1 + 1*3^0$	28	Ternary
1001	$1*4^3 + 0*4^2 + 0*4^1 + 1*4^0$	65	Quaternary

Figure-10 Representation Changes the Meanings

The slit experiment therefore has a close analogy with the process of perception. The perception of color for instance involves a two-staged process involving the eyes and the brain. The human eyes *digitize* light into the red, green and blue color types. The brain receives a signal that contains different proportions of red, green and blue. The slits in the quantum experiment are analogous to eyes that digitize the color prior to representation. While human eyes digitize colors in three types, it is possible to imagine that there could be other representation schemes that use two, four or seven colors to encode the signal to the brain. The battery of detectors in the quantum experiment is analogous to the brain that receives the color information as a representation in terms of ingredient colors.

The meaning of a position on the detectors cannot be given independent of the number of slits. Figure-10 illustrates different values that the digits 1001 will denote in binary, ternary and quaternary representation schemes. This idea can be extended to type distinctions in the case of quanta. The position of a quantum on the battery of detectors only corresponds to the *digits* but does not indicate the *value.* By changing the number of slits, the quanta will arrive at different locations on the battery. The information in the quantum is not changed by the number of slits, although its representation is. Information in the symbol corresponds to the value and not to the digits and the value is to be computed by converting the digits into a number based on the representation scheme. Similarly, the position probabilities in current quantum theory are insufficient without considering the number of slits used in the experiment. This problem is further illustrated by the fact that the same wavefunction can be represented in terms of different bases, each of which predicts different position outcomes. Which of these position results in a true description of the experiment?

These problems are solved in SI because reality in quantum theory is information, which can be represented using different encoding schemes. Different eigenfunction bases pertain to the representations. Since quantum theory does not connect the bases to the number of slits, it does not predict which basis corresponds to which experiment. Quantum theory only says that all experiments will be probabilistic, although the probabilities and outcomes in each experiment are different. The semantic approach is empirically testable because it can distinguish between experiments.

In a weighing scale experiment, there is a difference between the standard weight used in measurement and the pointer that gives the numerical result of the measurement. In current quantum theory, the battery of detectors is both the scale of measurement as well as a pointer. Thus, we suppose that if a quantum is detected at a detector, it is because the detector fired, which is because it detected a quantum. This assumption is not needed if the 2-slit is the standard for measurement and the battery of detectors is the pointer movement. Now, the detector represents the result of the measurement against a standard, which in this case is the choice of two slits. This partially addresses the problem of probability, because the obvious conclusion now is that if the experiment gave a different value result, measured against a standard, then that difference must be due to a difference in the properties of the measured object.

So, if the information in the objects remains unchanged, the detectors that click with a higher number of slits would be closer as compared to the lower number of slits. This is because as we increase the number of slits, the same numerical value (the classical position of the quanta) represents different information. The detectors in this case are simply pointers of a measurement device that swing depending on the standard that sits in counterbalance to measure the information in a quantum object. If the standard is large, the needle swings less and if the standard is small, the needle swings more.

The Meaning of Quanta

Part of the reason why quantum theory is problematic is that the idea of quanta itself isn't very clear. Why is matter not infinitely divisible? What physical intuitions underlie quantum discreteness?

The answer to this problem in SI is that the atomicity of things comes from the atomicity of knowledge and actions. Atomicity of knowledge means that there is a limit to the smallest or the most elementary ideas we can think of, represent and communicate. Atomicity of action means that there is a limit to the smallest operation or action that we can perform. When we start analyzing concepts and actions, we can reduce them to some elementary concepts and actions. But beyond a point, we will arrive at axioms and operators that are the simplest things we can conceive of and execute as actions. A further division of concepts and actions is impossible, when the axioms represent the most basic ideas that we can think of and operators represent the most basic actions we can perform. All knowledge is built up of elementary alphabets and all formulae employ basic logical operations like addition, subtraction, or multiplication. Knowledge and programs cannot be reduced indefinitely, and quanta represent this limit to reduction.

The atomicity of matter points to atomicity in our methods of knowing and doing. But these aren't just human limitations. They are also limits of logic and conceivability in nature. Knowledge is built by applying logical operations upon axioms. An axiom is an indivisible unit of knowledge and a logical operator is an indivisible action. Our knowledge is based on elementary ideas and procedures are based upon elementary operations. These elementary ideas and actions can be combined to construct complex ideas and operations. But the elementary ideas and operations can't be divided because they are elementary. If our methods of doing and knowing require us to perform atomic operations and think atomic ideas, then the result of this thinking and doing must obviously reflect in the smallest objects we can think of and create. The atomicity of matter is not independent of our methods of knowing and doing. Atomicity in fact shows the smallest concepts we can think of and the smallest operations we can perform. Quanta represent the limits of thinking and doing in nature, the limits of how small we can think and act.

A quantum is a unit of information (concept and/or formula). This unit can be large or small. Large concepts and formulae can be analyzed into smaller concepts and formulae, and there is a limit to how small a concept or a formula can be. However, quantum theory isn't just about atomic units of information. Quantum theory is about information per se—which can be atomic or macroscopic. Quantum theory is a fundamental theory of nature not because it deals only with 'fundamental particles'—the smallest constituents. It is a fundamental theory of nature because information is the basic constituent of everything else that makes up the physical universe.

By changing an object, we don't just change its physical properties. We also change the information it holds. The latter represents the concept used to describe the object as well as the laws that govern its behaviors. Changes to a symbolic object therefore aren't just quantitative changes to its physical properties. They are also theoretical changes to the laws that govern that behavior and conceptual changes to ideas in terms of which we describe nature. In effect, changes to objects involve *programming* an object whereby the object becomes a different concept and formula. A quantum is not a billiard ball whose motion changes quantitative values but leaves laws of motion and type intact. A quantum object is a computer, whose behavior is governed by the meanings and programs represented by its physical states, and which becomes a different type of computer if the data and programs in it are changed. Without changing aggregate material properties, a quantum object can be reconfigured by loading a new type of program and data into it.

The collapse of causality in quantum theory is a vista to a new science in which matter is programmable. The behavior of a quantum is not driven by deterministic laws of motion—the kind that Newton formulated for classical mechanics. The behavior depends on how the object has been programmed. Quantum theory needs to describe how matter can be programmed with concepts and programs, or how material objects can become symbols of concepts and formulae. Quanta are extended objects, and current quantum theory reduces these objects to classically measurable values. If the extended objects were instead viewed as forms associated with meaning—like we associate shapes with meanings in ordinary written language—physical properties could also be described semantically.

The Meaning of the Wavefunction

The standard Copenhagen Interpretation (CI) claims that matter exists in a superposed state. Unlike the Ensemble Interpretation

(EI) where a wavefunction describes a collection of objects, in CI, the wavefunction describes an individual object. However, the individual object in CI is in a superposed state, which means that reality itself is not in a definite state for us to know it. The definite state is created at measurement, and hence quantum theory—although a final theory of reality—will forever be probabilistic. In EI, quantum theory is not forever probabilistic; it is probabilistic only as far as we don't know how to describe the individual quantum and the present theory is a stepping stone to a better theory that overcomes incompleteness. The current theory is a description of a collection of quanta but not the individual quantum in EI. So, if there may be a final theory, what will the wavefunction represent in that final theory?

In SI, atomic objects are information-carrying symbols. The collection of symbols is also a symbol although its information is greater than the information in the individual symbols. The individual symbols are just physical properties, but they acquire conceptual and programmatic meanings in an ensemble. The dynamical properties of the quanta represent this information. Thus, a wavefunction represents an information bearing proposition. The wavefunction is not a superposition of states. It is also not a collection of objects that cannot be described individually. Instead, the wavefunction is a *macroscopic* quantum being described in terms of its constituents, like the meaning of a sentence can be expressed by word probabilities. Quantum theory does not need a collapse, because the wavefunction is already a macroscopic object, although these objects don't behave classically. The wavefunction represents an ensemble, not a superposition. It is a *combination* of constituents, which is currently described in terms of the constituents, probabilistically. When quanta are interpreted as meanings, a wavefunction will not represent an atomic object, or a collection of atomic objects. It will rather represent a macroscopic object.

The everyday world consists of macroscopic objects built from atomic objects. In everyday language, macroscopic objects are described conceptually—cars, houses, airplanes, people, etc. Everyday language does not reduce macroscopic objects into atomic objects to describe their properties. If quanta are treated as symbols of information, the distinction between atomic and macroscopic can also be dropped in science. A large object will denote complex meaning and a small object will denote simple meaning.

Think of a vegetable soup made of vegetables, spices, herbs, salt, milk, corn flour and water. Each of these ingredients can be individually tasted outside the soup. When we cook the soup, we use a recipe in which the ingredients are added one after another. Cooking the soup is state preparation because a procedure is followed to prepare a system prior to measurement. Once the soup is prepared, the ingredients exist in a *combination*, not a *superposition*. The difference between the two is that in a combination we can taste the *soup* whereas in a superposition we can only taste vegetables, spices, herbs, salt, milk, corn flour and water. However, we must describe the preparation of the soup in terms of its constituents. Quantum measurements correspond to a description of the soup in terms of its constituents while the quantum wavefunction represents the combination of ingredients forming the soup. Current quantum theory refers to the ingredients, but not to the soup itself.

In quantum theory, the combination corresponds to knowing the entire wavefunction as a certain kind of object that has been obtained by combining individual objects. The combination of individual objects creates a macroscopic object like we would create a sentence by combining words. The sentence has a more complex meaning than the words which constitute it. The meaning of a sentence goes beyond the meaning of its constituent words because the same words can be combined in different ways and yield a different meaning. Just as it is possible to know that something is a car, house or airplane without knowing the atoms that comprise it, the information depicted by a wavefunction can be described without knowing the constituents that compose it. Current theory reduces the wavefunction to atomic objects. If there was a method to combine atoms into macroscopic objects, this wouldn't be necessary. This possibility already exists in the present theory. Take for instance the wavefunction of an atom such as Oxygen. The wavefunction of Oxygen atoms does not represent a superposition of electrons and protons. It instead describes a *combination* of sub-atomic particles. The particles in an atom's wavefunction represent different electron *orbitals*. The atom is a combination rather than a superposition of those orbitals. In effect, an atom is a complex object as compared to the sub-atomic particles, but it is a distinct object rather than a superposition. And yet, we can divide it into its constituents.

Viewing the quantum wavefunction as a macroscopic object is important because it means that quantum theory can be applied to macroscopic objects, and not just sub-atomic particles. Quantum theory is about information-carrying objects, whether these objects are atomic or macroscopic. State preparations and measurements must employ an understanding of how information can be encoded and decoded. Quantum theory is incomplete because it does not explain how information can be encoded or decoded. Adding information to quantum theory requires two things—(a) recognizing that quanta are symbols and (b) quantum theory is enhanced to treat a symbol in three ways as things, meanings and formulae.

Eigenfunction Orthogonality

Hilbert Space describes matter as orthogonal alternatives. These alternatives must be treated as distinct objects to avoid the Copenhagen conclusion that the same object is in multiple states at once. In SI, the components of a wavefunction are distinct symbols and orthogonality refers to their mutually exclusive meanings. While this solves the issue of quantum probabilities—probabilities are a feature of a collection of objects—it still does not explain why quantum reality must be described as a collection of *distinct* objects. After all, when we form a collection, the individual objects have identical mass, charge, energy etc. The distinction between objects in a collection is not known at the time of state preparation and not created by the experimenter at that point. How is a collection of physically identical objects differentiated inside a collection when state preparation does not involve acts of differentiating?

This problem can be understood with an example. Consider how people structure themselves into teams. Prior to their coming together, everyone has some skills, experiences and behaviors. But within a group, the individuals must evolve into a cohesive unit. That happens when everyone takes a different *role* in the group. Someone must become a coordinator; another person may create strategies;

a few people will execute those strategies while others will remain silent observers or offer new ideas. The role played by everyone makes them distinct in the group. This distinction isn't necessarily the way in which people were distinct outside the group. To stay as a unit, individuals must organize themselves into orthogonal functions and roles, because that makes every individual a distinct type. By eliminating conflict and role overlap and identifying a unique and distinct part that everyone can play, the group is unified.

This example shows a vital difference between classical and everyday collections. A classical particle's identity is independent of the collection. A quantum particle's identity is tied to the collection. A classical particle can be anywhere in a collection. A quantum must be at a definite location within the collection. A collection draws a boundary between what is inside and outside although classical physics has no such distinction. In classical physics, particles move according to forces and the term 'collection' has no significance, because if we described all the individuals we would have also correctly described the collection of those individuals. In quantum physics, the collection is vital to the description of individuals.

The reductionist stance of classical physics is false in quantum theory. Now, the very notion of an individual object is defined by its role in the collection. The collection still reduces to individuals, but the individuality of the individuals is defined by the collection. The orthogonality of alternatives in the wavefunction represents the fact that an individual quantum is defined by giving each object a unique and non-overlapping identity in a collection. The identity of an object is not defined independent of the collection as in classical physics. Rather, as people organize themselves into orthogonal roles within a team, quantum objects automatically order themselves into a set of orthogonal alternatives within a collection. In this sense, quantum collections also show properties of everyday collections.

Quantum theory is *linear*—and the whole is a sum of its parts. However, quantum theory is not linear like classical physics where the parts are independent of the whole. In classical physics, a particle is unchanged outside and inside the collection. In fact, boundaries are only a theoretical convenience used to eliminate objects outside the boundary from the causal picture. The linearity of quantum theory

requires us to define a quantum in relation to the whole. The boundary is paramount to identifying which objects constitute the whole.
The linearity of the quantum object must be compared to a jigsaw
puzzle where the whole reduces to the parts, but the parts depend
upon the whole picture that is being reduced to parts.

A quantum wavefunction can be represented in multiple bases, just
as a picture can create multiple jigsaw puzzles. Each puzzle linearly
decomposes the whole picture into parts. The parts are empirical.
The reduction is not just a theoretical convenience but corresponds
to measured objectivity. The different puzzles have their origin in
changes to the experimental setup by which reality will be known in
different ways. In the interference experiment, this can be achieved by
altering the number of slits in the experiment. In each case, the same
macroscopic reality is decomposed into parts differently. Quantum
theory creates the possibility for the *interpretation* of reality where
the same macroscopic object is divided into different atomic objects.
The experiment tells how reality is composed of *logically* distinct and
orthogonal parts, although there are many such ways to know the
world.

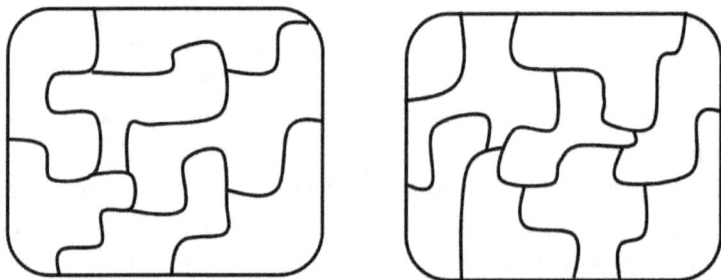

Figure-12 The Quantum Wavefunction and Jigsaw Puzzles

The quantum descriptive approach is like the use of axioms in mathematics. Mathematical theories choose one set of axioms for a theory,
but the theory can be based equally well upon other sets of axioms.
The selection of an axiom set is ultimately a choice, which makes no
difference to the theorems (statements) constructed in the theory, but
it makes a huge difference to the meanings we regard fundamental.

The reduction of the wavefunction into orthogonal alternatives is also a choice in which reality is described differently by altering the relation between the measured and the measuring systems. This choice does not change the macroscopic wavefunction (counterparts of theorems in a mathematical theory), but it changes the constituents of that wavefunction. These constituents are the axioms or alphabets of the description. By changing the axioms, we change the *language* in terms of which nature is described. Effectively, we are describing nature in terms of different concepts, symbols, axioms, and while each language produces the same meaning, it changes what we regard as "atomic" in each case.

Quantum vs. Classical Statistics

There are two ways in which probabilities in quantum theory can be thought of: as features of individual particles or those of collections. In the case of a coin toss, if the coin is not biased, there is a 50% chance that it turns up heads or tails. This is a feature of the individual coin, because in another instance the same coin would produce the opposite result. Contrast this with a case where a ball is drawn from a bag of black and white balls. If the number of black and white balls is equal, then there is a 50% chance that a black or white ball will be drawn from the bag. However, now, the probability is a feature of the collection and not of the individual ball. In the case of a coin, the *same* coin produces heads and tails. In the case of the bag, *different* balls are white or black. This analogy is important because it shows that quantum probabilities can also be seen in two ways.

If probabilities are applied to the individual quantum then the eigenfunctions in the wavefunction are states of the individual quantum. The *same* object may be observed in state $S1$, $S2$ ... SN. However, if the probability is applied to the collection of quantum objects, then $S1$, $S2$ ... SN pertain to different objects and not states of the same object. In this case, a quantum is always in a definite state, but by repeating the experiment many times, different quanta will be observed in different states. Since the individual quantum is not uncertain there is no statistical problem as far as the individual quantum is concerned.

There is, however, an uncertainty about which quantum turns up when a measurement is performed. Contrast this with the case where the probability is applied to the individual object and there is a problem in describing the individual object's state and hence the state of the particle ensemble as well.

In SI, probabilities are applied to the ensemble and not to the individual quantum. The quantum experiment is the act of drawing a ball from of a bag rather than the tossing of a coin. A coin cannot be described completely in terms of the facets because the facets cannot be observed at once. In the case of the colored balls each ball has a single color and the balls can be known both individually and collectively. In the 2-slit experiment, for instance, we know the individual detections if we draw the quanta one after another. We also know them collectively if we view all the detections over time. The colored ball view of probability is a better way to interpret quantum probabilities because it makes both the whole and the part knowable and empirical. In the case of the coin, the whole is knowable but not empirical, since all faces cannot be seen at once.

With this treatment of probability, the wavefunction doesn't per se describe the *state* of anything. It is certainly not the state of the individual quantum because the quantum is in a definite state and not in a superposed state. It also doesn't describe the state of the ensemble because the ensemble is a combination of the parts and is therefore not in a superposed state. The notion that the wavefunction describes a state inevitably leads to the collapse problem because the whole must be reduced to one of the parts. The wavefunction is simply a function that describes (a) the logically distinct components of a whole and (b) the relative probabilities of finding a particular type of logical component if we happen to make a draw.

Quantum probabilities are also distinct from classical statistical states (especially as used in statistical mechanics) because the statistical mechanical state is a state of both the individual and the whole system[1]. The state of the whole system in statistical mechanics has no counterpart in quantum theory because the state of the whole system is never described in quantum theory. The probability associated with the whole state in statistical mechanics is therefore different from the probability associated with parts in quantum theory. The total number of states in statistical mechanics is physically significant (it is related to the system's entropy), but the total number of states is irrelevant in quantum theory. Having more states doesn't mean the quantum system is more chaotic. Having more eigenfunctions means that the information system has a richer vocabulary with a greater number of concepts or axioms.

Epistemic Atomism

This brings us to the question of what a quantum is. Following the ensemble view, the wavefunction is obviously not a quantum, because it represents a collection of quanta. Individual quanta are the orthogonal eigenfunctions in the wavefunction. And because these eigenfunctions can be changed from one basis to another (when the experimental setup is modified) the quanta do not represent metaphysical atoms anymore. The idea of metaphysical atoms goes back to Greek times where Democritus postulated that matter is comprised of atoms of Earth, Water, Fire and Air. These atoms were supposed to be unchanging entities. Macroscopic objects combined these atoms, but there was no way to alter the identity of the atoms. I shall call the Greek view *metaphysical atomism* since it describes atoms as things that exist independent of whether we know them.

With quantum theory, the division of a macroscopic object into atomic entities is subject to observational conditions. When reality is observed, it is cut up into atoms according to the experimental setup. In each such cutting up, reality is described using a different language, because it has been divided into words and concepts unique to the experiment. Reality therefore is not an *a priori* fixed set of atoms. Rather, atomicity depends on the experiment that decides the language used to describe nature. By changing the language, the experiment changes what is axiomatic and what is seen as atoms. Unlike the metaphysical atomism of the Greeks, atoms in quantum theory aren't the fundamental things that exist independent of the observers who know them. Atoms instead represent the basic axioms or words of language in terms of which nature is described and this language can change based on experiments. So, what is regarded fundamental in one situation may not be so in another situation. Each experiment can invent its own *vocabulary* to describe the reality within that experiment. The vocabulary is not fixed *a priori*.

This flexibility in dividing nature into parts implies that any macroscopic object can be divided in many ways. The encoding and decoding procedures can use different languages. The senders and receivers of such semantic messages will be able to communicate because they obtain the same macroscopic information, although they will differ in the languages they use to write and read such messages. In effect, quantum observation represents a translation between languages like a meaning that may be expressed in different languages like English and German. In each language, different axioms, words, or concepts are regarded fundamental and the description is constructed in terms of those presumed fundamental ideas. Quantum theory thus supports the possibility that a system is encoded in one language but decoded in another language. The information content across such encoding and decoding shall remain unchanged even though the axiomatic basis is being transformed.

Atoms are the smallest ingredients of our knowledge and action. An atom is a concept in terms of which reality is understood and changes to it are described. Each field of knowledge regards different concepts as fundamental. For a sociologist, the unit of knowing is a human being. For a biologist, the unit of knowing is a cell. Molecules are units of knowing for chemists and sub-atomic particles are units of knowing for physicists. The concepts regarded as fundamental depend on the context. A quantum eigenfunction is created by dividing nature under an experimental setup. This setup can be likened to different observational stances such as between biologists and sociologists, because they will partition nature by different concepts and to different levels of granularity. Each concept represents a type of object associated with a type of behavior. Essentially, the output of an experiment depends on the way matter is divided into parts. Nature can be known in many ways corresponding to the different *interpretational stances* associated with different types of languages. The language used in science predetermines the methods of knowing and changing the world.

This can be called *epistemic atomism* as quanta represent the smallest elements of knowing or action in a given context. In epistemic atomism, the smallest that exists is the smallest that can be known and used, different from metaphysical atomism, where the smallest that can be known or used is the smallest that exists. By refining our concepts and operators, we can divide nature into finer parts, and consequently know nature more atomically. But there is a limit to how small we can think and act, and this limit is the smallest things that can ever be thought of and done. If what we can know is that which exists, then there is no choice in knowing. The universe and its knowledge are fixed. To know reality, we must know it in the finest possible manner. If, however, what exists is what we can know, then there is choice in describing nature in many ways—all compatible with quantum theory as a theory of nature. As schemes of our knowing change, the reality that we know is adjusted.

Quantum theory is not a definite description of how things exist independent of our methods of knowing and doing. The theory instead is a *framework* to describe the interaction between the experiment and its observations. By changing the number of slits in the slit experiment, the observer can change the outcome of the measurements. The theory is compatible with an infinite number of experimental representations (or descriptions) of the same reality. The eigenfunctions in a wavefunction represent the result of dividing nature into mutually exclusive and logically elementary parts.

A theorist can choose to not describe the universe as electrons, but as more complex concepts. Chemistry, biology and physiology already describe nature in terms of complex constructs, although these are regarded fundamental concepts in these subjects. As scientific descriptions probe deeper into nature, they will search for more fundamental things. The "quanta" will now become smaller. The term quantum therefore represents the *axioms* of any theory and not metaphysical things. If the axioms are fundamental, then nature is described in terms of fundamental things. If axioms are more complex, then the description is in terms of complex constructs. In both cases, the descriptions use certain conceptual "atoms," or what a scientist considers conceptually fundamental or axiomatic.

This should however not suggest that fundamental concepts are always arbitrary. If this was the case, then two individuals who viewed the world differently could never communicate. Nature allows flexibility in interpreting a macroscopic object in terms of many atomic languages, but these languages are connected by a transform, and are not arbitrary. The atoms in the language co-vary in such a way that the total meaning in the macroscopic object is preserved. Different interpretations are translations of a meaning in different languages like we might express a meaning in English or German. The atoms in the description change, but the macroscopic meaning is unchanged. Since all quantum phenomena arise in ensembles, the flexibility to alter the representation scheme without altering the overall meaning exists everywhere. An observer can translate meaning encoded in one language into another language's representation. But an ensemble cannot be interpreted as arbitrarily. With an arbitrary meaning, science too will be arbitrary.

The existence of many possible languages presents a problem: Which language must science use? There are two ways in which this problem can be addressed. First, we might choose one language and call it the language of scientists. Every scientist must express their observations in this language and scientists will be able to understand each other. But, of course, there can be other languages to describe nature as well. Each group of scientists that chooses a different language will create a sub-culture that describes nature differently, although they can communicate. Since science is not just about the communication of observations but also an agreement on the fundamental nature of reality, this is somewhat unsatisfactory.

Second, it is possible that all languages are mutations of a universal natural language. To describe nature, we can discover nature's language. This language can represent the limits to the division of nature into parts. Note that when we employ different languages, some concepts must be 'bigger' as compared to concepts in other languages. For instance, the German word *Weltanschauung* and the English word *worldview* describe complex concepts that can be reduced to simpler notions. If there is a limit to divisibility in nature, then there is indeed a fundamental set of indivisible ideas and actions, although individual languages may not use them. The limit represents not just the smallest things that metaphysically *exist*, but the logically most fundamental things that can be known and done. To find the limit to reduction, science must find limits to thinking and doing. The smallest things in nature are the most elementary ideas and formulas that can exist. Science can agree to use this basic language as it constitutes the limits of thinking and doing.

Obviously, the second alternative is better because it simplifies the communication of scientific results. If there is a limit to reduction in nature, then different languages can be reduced to one fundamental language. In case of a dispute about the fundamental nature of reality, information can be converted to the one that uses the most fundamental concepts. Note how a single universal language is not needed for the communication of macroscopic information. Several languages can do the job equally well. A fundamental language is needed to understand the *limits* of divisibility in nature. The fundamental language serves to communicate macroscopic information and the nature of atomic reality. In that sense, the same language can serve the purposes of everyday and scientific communications. Different sub-cultures can understand each other by translating their descriptions into a language that uses the most fundamental concepts. Now, sub-cultures are derived from a universal language of nature, and while nature can be known variously through multiple languages, there is still a universal view about nature. This view describes things as they are, because every possible description can be reduced to this universal language. This universalism can assert a metaphysical reality, without precluding the existence of many individual realities. Every language creates its reality and in so far as science is the empirical confirmation of reality, all descriptions are real. However, there is still a description that is the "ultimate universal description" because it is made in a language that every observer can potentially understand and use.

Quantum theory can be a theory about that universal language. As science discovers more and more elementary ideas in human thinking, science will also be able to break nature down into those logically elementary pieces. There however comes a point at which science has reached the most elementary *thinkables* and *doables*. The reduction of nature must stop at this point because science has now reached the set of most atomic ideas and operations that can be thought and practically used. The ultimate quanta of reality are the most elementary things that can exist, can be known and used.

If a symbol can represent existence, knowledge and action, then the most elementary things that exist are also the most elementary things that can be known and used. A symbolic view of reality avoids the problems of knowledge where what exists in nature may not be knowable or what exists and is knowable may not be usable as technology. The limits of language are the limits of things that can exist, can be known in science and can be used as technology. The limits to science are limits to scientific language which prescribes the things that can be thought of, used by actions, and will physically exist. Stretching these limits in language will stretch the possibilities in nature. However, this stretching does have its limits.

Quantum Complementarity

Complementarity was added to quantum theory by Bohr to suggest that position and momentum are two complementary ways of describing an object's state. Complementarity has been subject to much debate, mainly because in quantum theory position and momentum are not different aspects of a complete state but different ways of *representing* the *same* state. In classical physics two variables $\{x, p\}$ represent two independent quantities that completely determine a state. Knowing the value of position x in classical physics doesn't determine the value of momentum p because both x and p can vary independently. In quantum theory, x and p are representations of the *same* state and are not independent. That is, knowing the position representation of a wavefunction fixes the momentum representation. The position and momentum eigenfunctions are related through a Fourier transform. The Fourier transform of a position eigenfunction results in many momentum eigenfunctions and vice versa. Therefore, knowing the position state does not fix the momentum state, and vice versa. This is intriguing because states pertain to the *same* object. How can an object have a single fixed position but many possible momenta and vice versa? Stated in this way, complementarity is often confused with uncertainty, and the theorist now argues that the theory forbids the simultaneous determination of position and momentum as in classical physics.

Recent quantum experiments[2] have invalidated the classical view of quantum uncertainty since these experiments show that it is possible to measure position and momentum to levels of accuracy not allowed by the uncertainty relations. These experiments show that we can make the state certain through state preparation. We cannot therefore interpret the uncertainty to mean that position and momentum are not simultaneously defined (as Bohr held) or that they cannot be simultaneously measured (as the dominant view of uncertainty relations claims). We need a new way of looking at uncertainty that steers clear of the above-mentioned notions.

The Fourier relation between position and momentum tells us that if the position has been fixed, the possibilities for momentum have been *constrained*, although not fixed. So, a particle could not be in an arbitrary momentum state if the position state is fixed. And the particle could not be put into any arbitrary position state if the momentum state has been fixed. This is novel in quantum theory because in classical physics fixing the position did not constrain the momentum state or vice versa. So, while position and momentum are distinct variables and they must be both defined simultaneously to specify the quantum state, fixing one of them partially constrains the other. In that sense, position and momentum are not *independent* variables as in classical physics. To understand this problem, we need some ideas that resonate with this aspect of the quanta.

Everyday objects are described in two complementary ways as *concept* and *use*; something can be called a 'table' and it is used to 'dine'; something is called a 'car' and it is used for 'driving', etc. Note, however, that the relation between tables and dining is not fixed. A sofa may also be used for dining, while a table may be used for studying. Similarly, a car may also be used to live in, while a truck may be used for driving. Not everything, however, can be used for every purpose. For instance, we cannot drive a table, and we cannot eat a car. While there is a many-to-many relationship between object-concepts and action-concepts, the relationships are not arbitrary. The conceptual form of an object restricts its possible uses, without singling out just one use. Likewise, the usability of an object restricts possible concepts we can attach to it, without singling out a particular concept. The complementarity between concept and use in everyday objects can be the basis for understanding the complementarity between position and momentum in quantum theory.

Now, we can say that the position eigenfunction represents object-concepts and the momentum eigenfunction represents action-concepts. Knowing an object-concept limits possible action-concepts but does not fix the action-concept. Similarly, knowing an action-concept limits possible object-concepts, but does not fix the object-concept. However, this limitation is not an enduring feature of reality. There can be, for instance, a car that can only be used for driving but not living (if the car did not have a roof on top). Or, a piece of cloth may only be used for wiping floors and not for protecting someone from bitter cold (if it did not prevent air passing through the cloth). These everyday examples suggest that while an object is described both by object-concepts and by action-concepts, knowing one of them does not fix the other. Rather, we require both modes of knowing to completely know an object and its behavior. Object and action concepts are therefore complementary and yet not independent. Knowing an object-concept (denoted by the position eigenfunction) does not fix the action-concept (denoted by the momentum eigenfunction), although it limits the possibilities. It is, however, possible that an eigenfunction has just one position and momentum. Indeed, electrons in a chemical atom have a single position and momentum state. These can be likened to the everyday notion of a chair that can only be used for sitting, and nothing else.

SI demystifies Bohr's complementarity as two ways of describing an object, as a concept and as a program. Knowing the concept limits possible programs but does not single out one program. A word processor can for instance be implemented in many ways, and some word processors may allow the ability to draw pictures while others only deal with text. In general, therefore, if we speak of a word processor, the notion is ambiguous unless we describe all its behaviors. By itself, the notion of a word processor does not single out all its behaviors. But this does not in any way suggest that a word processor itself will have ambiguous behavior. The ambiguity only pertains to our use of the term 'word processor' to describe something succinctly because that short-form description does not capture all the possible behaviors of the object. We could also write a long-form description such as "a piece of software that helps you write, edit and delete text, but does not include a spell-checker, picture editor or text formatter" and such a description would limit the possible uses of the program. In the limiting case, when an object has an atomic concept attached to it, it will only permit one operation. As the concept becomes more elaborate, it will permit many possible uses. For instance, if the word processor included picture editing and text formatting and styling, it could be used to create posters and flyers, and not just entering plain text.

Uncertainty Relations

While complementarity is a fact about the wavefunction, the uncertainty relations pertain to measurements on a wavefunction. The uncertainty relations state that position and momentum operators do not commute, and we can't define position and momentum simultaneously. A quantum object either has a definite position or a definite momentum but not both at once. To understand uncertainty relations, we must recognize that quantum objects are always extended. Think of an ordinary classical object like a chair. The chair has a definite position, but we can't point to that definite position, because the chair exists in multiple locations at once due to its extension in space. In classical physics, we reduce the position of the chair to the position of its center of mass. Such a reduction will however rob the chair of what makes it a chair—namely its form and function. As point particles, the objects cease to have conceptual names and properties. By reducing an object to its center of mass we will deprive the object of what makes it unique. That is the reason why quantum objects can't be reduced to points and quantum extension is irreducible because the form of the quantum denotes a specific type of object, not just the spread of matter in space.

If objects are fundamentally extended, we also cannot define a specific momentum for the entire object because the particle's state of motion may be defined by a changing phase within a static state. Different points within an extended particle may also have opposite phases. If we sum the phases in a stationary state, the particle may not be moving at all, because the particle is in a stationary state. Particle extension therefore also makes momentum uncertain not because the particle lacks a definite momentum but because it is impossible to define a unique vector for the particle. This is like the problem in determining the motion in ordinary classical extended objects where different parts are moving in different directions. It is easily visualized in the case of vibrating membranes where different parts of a membrane move in different directions and speeds.

In fact, we should stop thinking about position and momentum in a classical sense—i.e. as the properties of point particles. We must rather think of the position and momentum *state* of the quantum. This requires a rethink on the nature of position and momentum. We spoke earlier about how positions are defined in an everyday sense as countries, states, cities, streets, houses, etc., which are all extended locations and yet we think of them as precise locations. Similarly, we saw how time can be divided into years, months, weeks, days, hours, etc., which are all extended durations and yet we think of them as precise times. This kind of thinking comes from a *hierarchical* view of space and time. The higher location in the hierarchy represents an extended location or time, but it is still a precise location quite like a country or a year are precise definitions. But if we try to reduce them to classical physical properties, we will find them uncertain.

Heisenberg's uncertainty relations specify the *lowest* bound on the uncertainty as $\hbar/2$. The uncertainty of a quantum can be greater, and this limit denotes the smallest amount of uncertainty corresponding to the smallest particle. The uncertainty associated with a location like a city (as a location) or a year (as a duration) is much bigger. But even the smallest particles will have some uncertainty, which indicates a limit to the divisibility in space and time. We might think that this uncertain position and momentum are classically problematic, and they are. Which is why we need to stop thinking of space and time in terms of point locations and start thinking of them hierarchically as more uncertain locations and durations. This uncertainty is classical, but the quantum state is certain. That is, the city or a year is indeed a real entity, not merely something we construct by combining the smaller entities.

The idea of a quantum essentially indicates that to store some information we need a finite amount of space-time. As the amount of information grows, the space-time required is higher. We cannot store an infinite amount of information in a finite amount of space. If an infinite amount of information can be packed in an infinitesimal space, it leads to singularity problems in classical physics where the self-force of an object grows rapidly as the information becomes more and more dense. Since a classical particle is a point, even a small amount of mass or charge will lead to the singularity problem.

In current interpretations, the uncertainty principle has two aspects: (a) by default a quantum particle's state is uncertain, and (b) we can never make this state certain. Experiments now show that (b) is false, while (a) is still true. This essentially confirms the idea that nature can be encoded with information. If nature hasn't been coded with information, then its state is uncertain. By adding information, we can compensate that uncertainty to create more certainty. However, we can never attain zero uncertainty in the classical sense because we can never divide space and time infinitely. That limit on divisibility of space and time is given by Heisenberg's uncertainty. Since the uncertainty can be greater than $\hbar/2$, nature is like a slate waiting to be written upon. If we read that slate before writing on it, then we will obtain greater uncertainty. But, if we have written on the slate, uncertainty is reduced to the extent of information encoding.

Schrödinger's Equation

Schrödinger's equation describes the evolution of the quantum wavefunction. It tells us that after we have put the system into a definite state, the system will change state, if the energy of the system is changing. The right side of the equation defines the changing energy while the left side describes the change in the wavefunction.

$$i\hbar \, \frac{\partial \psi}{\partial t} = H\psi$$

However, since the wavefunction represents probabilities of quantum states, this evolution is different from the evolution of states in classical physics where given initial states of each particle we could predict their final states. In quantum theory we can only predict the altered probabilities. Change in energy in a quantum system represents information change over time, but the current theory only predicts how one symbol-set will evolve into another in the aggregate. Schrödinger's equation says that the vocabulary in which we describe a system will evolve deterministically, but the information coded using that vocabulary is unpredictable.

Present quantum theory describes the evolution of vocabularies, but not the evolution of propositions created from these vocabularies. If particles are symbols and their states represent information, then it should be possible to describe the evolution of propositions along with the evolution of vocabularies. This is not possible in current quantum theory. In a semantic reformulation of quantum theory, a new equation would be required that describes how information encoded by the ensemble evolves. Such an equation would be able to predict the evolution of an ensemble's meaning.

In a semantic view, wavefunction probabilities are indicators of symbol frequency in a proposition. This is an incomplete description of the ensemble state, because the ensemble represents a proposition that sequences the symbols. These sequences can be observed during measurement, but they cannot be predicted by current quantum theory because the theory treats the quanta as particles and not as symbols. The order of the symbols is empirical, but it is not predicted by quantum theory. When the symbol order can be predicted, then the dynamical equation that replaces Schrödinger's equation will represent the evolution of information rather than the evolution of vocabulary. Even without such an equation, we can see that semantic ideas lead us to the view that the wavefunction represents a collection of types—i.e. symbols in a vocabulary. Even the current Schrödinger's equation can therefore be viewed as representing the evolution of vocabularies by which we can guess that the information content in the system is changing, although we can't predict the exact type of content because we can't know the *order* amongst the individual particles states in the ensemble.

In classical physics, we speak of how a particle moves from state S1 to S2 while preserving its identity. Thus, we say that the particle at states S1 and S2 is the *same* particle, and it is possible to describe the state evolution of every particle. In quantum theory, we cannot say that the evolution of the ensemble is the evolution of the *same* particles. This is because, while particle identities are fixed in classical physics, their identities are fungible in quantum theory. Every ensemble consists of multiple types of objects. The evolution of the ensemble is the evolution of one type-set of particles into another type-set. This evolution is not identity preserving, and thus the identities of objects aren't preserved in time. When a quantum system evolves to a new wavefunction, the particle identities and types in that wavefunction are also different. One vocabulary and the order of these symbols will change into another vocabulary and order. Such transformations are not identity preserving and certainly not meaning preserving. These transformations are also not energy or matter preserving because energy or matter needs to be added or removed from the ensemble to change its wavefunction.

This means that the way we have envisioned determinism (and predictions in science) now needs to change. The determinism isn't at the level of individual objects, because a collection of objects will interact, and their identities will change in the process. Determinism should now be thought of at the level of collections. It is well known in classical physics that determinism at the individual particle level doesn't work even for classical ensembles because the ensemble does not follow reversible laws of classical mechanics. But the failure of determinism led to the impression that science can at best make probabilistic predictions. That conclusion is false in a new theory that describes an ensemble as a proposition rather than a collection. This evolution is predictable if particles are information symbols. But the prediction would require a theory that allows us to combine individual symbols into complex, macroscopic symbols. This must be a semantic theory. A meaning-encoding ensemble would evolve the knowledge represented in the ensemble over time. The ensemble can now be viewed as a computational system that computes the evolution of knowledge. If we encoded a proposition in the ensemble, the evolution would automatically compute a new proposition. This computation may not be vocabulary preserving and the new proposition may use a new vocabulary to describe new claims that logically follow from mutating the original proposition.

Understanding Non-Locality

Non-locality is the consequence of quantum entanglement or what Einstein called 'spooky action at a distance.' One example of non-locality is a two-particle entangled system with opposite spins. If these particles are drawn apart, and then spin measurement is separately performed on them, knowing the state of one of them allows us to *know* the state of the other. If, subsequently, a measurement is performed on the other particle, the measurement outcome is confirmed, and the particle is in the state predicted by the first measurement. How can one measurement predict the state of another particle? Classically speaking, if two particles are a distance apart, then, to fix the state of the second particle some causal communication (i.e. energy transfer) must take place from the location of the first measurement to the location of the second measurement. This causal communication must take a finite amount of time governed by the finite speed of light that determines the minimum time needed for any energy transfer. Quantum non-locality violates this principle. The second measurement, for instance, can be performed immediately after the first, not allowing enough time needed for an energy transfer. The faster-than-speed-of-light communication appears to violate the theory of relativity which postulated the speed of light as the fastest speed possible.

Non-locality appears to be a problem in CI, which claims that the wavefunction pertains to the state of the individual system and not to the ensemble. Therefore, when the wavefunction reduces into one of the eigenfunctions through the so-called collapse postulate, it appears that some causal agency must be involved in the collapse, which must travel between two locations faster than the speed of light. But this is a problem only in CI with its collapse postulate. In the ensemble view, the wavefunction describes a pair of particles and not the individual particles. A measurement would detect one of these particles although we can't tell which one upfront. But, once we have found one particle at one detector, we can know for certain the identity of the second one because there are only two of them.

In SI, the entanglement of the two particles represents a special case of a wavefunction in which there are only two particles. These can be represented as a pair of opposite bits—1 and 0—or a pair of conceptual opposites like hot and cold. Knowing that a detector finds a cold particle, we can be certain that the other will find the hot particle. That is not the same as *experiencing* the second alternative. We just *know* that the second would be the opposite of the first, but we don't experience the opposite result in the experiment. If, however, a measurement is performed on the second particle, we will observe the opposite result, just as our knowledge from the first predicts. This is because they are opposites and only two of them.

Non-local action at a distance is not necessary to fix the state of the second particle as the particles are not in superposed states of many alternatives. Each particle is already in a definite state, although we don't know which particle is in which state. The probabilities of the wavefunction don't represent an uncertainty about the *individual* particle state. They only represent the frequencies of occurrence of different particles in an *ensemble*. Non-locality appears to be a problem when the wavefunction probabilities are applied to the individual quantum and not to the ensemble.

Non-locality tells us something interesting about the nature of knowledge, namely that every logical distinction (1 and 0) is defined through a mutual opposition and not independently. We can't *define* the meaning of hot without saying that it is opposite to cold. Thus, hot and cold do not have an independent definition and their entanglement represents the fact that they are defined mutually by an opposition. This is not just a fact about binary distinctions but about all N-ary distinctions. The N eigenfunctions in a wavefunction represent an N-ary distinction in which each eigenfunction is conceptually orthogonal and distinct from others. All these eigenfunctions are mutually defined or not at all. The eigenfunctions collectively represent the *form* of an N-way distinction.

An ordinary example of this distinction can be seen in the seven notes of music that are defined through a mutual distinction. Individual frequencies don't make a note unless we define the entire scale of notes by defining the notes in a conceptually distinct manner. Each note has an associated physical property of frequency, but they are also semantically distinct as notes. By shifting the scale, we would shift the frequencies but not the distinction between the notes. That means that even in a shifted scale we can use the terminology of the same seven notes, and their interrelation will be unchanged.

A musical composition would not be meaningful if that composition only used one note. To make a meaningful composition, at least two (if not more) notes are needed. With a single elementary concept, we cannot create complex concepts. But with two or more elementary concepts, we can construct a wide variety of concepts. The meaning in these constructions arises not because of what the concept is, but also because of what it is not through a distinction with other concepts that are possible but not instantiated. Distinctions constitute the negative aspect of knowledge where we know something by knowing what it is not, in addition to what it is.

The Paradox of Change

In Schrödinger's equation, a wavefunction will change provided the Hamiltonian—which represents the total energy in a system—also changes. The energy can change if a force field is applied to add or remove energy from the system. However, this force field is also another quantum ensemble, which must be in a stationary state unless another force field is applied to change the state of that ensemble. Now it seems evident that to change the state of an ensemble we need another ensemble to change state first, which in turn needs yet another ensemble to change states, and so on, ad infinitum. Quantum state transitions embody a counterpart of the famous Zeno's paradox which argues that change is impossible. In Zeno's paradox, for an object to go from state A to B, it must go half-way through to that state first. But before it can go half-way, it must go one-fourth of the way, and so on, ad infinitum. In quantum theory, changes are discrete because quantum particles are discrete. So, an infinity of steps in carrying out a single change is not necessary. However, now, it is impossible to do even a single step unless energy changes, which in turn requires another system to change its energy. In Zeno's paradox, change is impossible because there is an infinity of steps *between* any two states. In quantum theory, change is impossible because there is an infinity of steps *before* a change.

A quantum state transition will take place *if* energy is added or removed. This addition or removal of energy is seen in quantum experiments that change the electromagnetic field surrounding the measured system. The changing electromagnetic field is, in turn, because some electric current flows through an apparatus, which is made to flow due to the flipping of an electric switch and because there is voltage potential in the electric wires, which are in turn due to other actions performed prior. Now, this was not a problem in classical physics where objects were always going to the next state because of their previous state. This is not the case in quantum theory because the quantum system settles into a *stationary state* once its energy is fixed. A classical system is always in a state of motion, and a universe once set into motion will always be in a state of motion. Quantum theory is different because it settles into a stationary state and does not automatically transfer its energy to other systems. A quantum system must therefore be *forced* to emit energy by the application of external fields while a classical system will do it *automatically*. It follows that a quantum system, if left to its own devices, will remain in its current state unlike a classical system which will change its state. If the universe is governed by quantum theory, then every system in the universe must remain in its current state. Then what causes the universe to constantly evolve?

The classical universe exists in a *single* space-time where every object can interact with every other object. Quantum theory, however, requires the universe to be divided into many ensembles, each of which is a closed region of space-time. If the entire universe was a single space-time, then according to quantum theory it must already be in a stationary state governed by Schrödinger's equation. If, however, the universe is divided into many ensembles, then each of these ensembles must by itself be in a stationary state. Whichever way we look at it, the universe must be in a stationary state. However, the universe is obviously not in a stationary state, which creates the following causality problem in quantum theory: What causes a system in a stationary state to emit or absorb energy? This problem cannot be solved by postulating that energy is transferred from one system to another because this transfer itself is unexplained.

In SI, two systems interact by transferring information. But the information transfer is not deterministic. It is rather governed by a *choice.* When a choice is made, then information is transferred, and its effect could be described as an energy transfer. Once the information transfer is complete, the receiver settles into a new stationary state. The received information does not automatically cause the receiver into transmitting information to other receivers again (although in the case of humans if the information was interesting, we would be inclined to share it with others).

The problem of missing causality in quantum theory requires a cause that interacts with information. I can only speculate here that this has something to do with the nature of the mind which comprehends meanings and makes choices. Certainly, the existence of choice is consistent with missing causality in quantum theory, because the theory predicts that the universe goes to a stationary state. To keep the universe in a state of change, it would seem, minds are needed to make choices about information transfer.

The gun in the hands of a shooter only describes how the gun *may* behave *if* the shooter pulls the trigger. The gun cannot fire on its own. To pull the trigger, there must be activity in the hands of the shooter, but that too is a quantum ensemble which cannot change its state unless the energy in the hand is changed. It follows that we must eventually trace causality back to some choices in the observer which are effected as changes in the total energy of a system. In classical physics, the force that causes the change is deterministic and in quantum theory that force should now represent a choice.

Matter and Mathematical Laws

In classical physics, the laws of nature have no physical existence, although they have a mathematical reality through logical consistency. Consistency however doesn't tell us how these laws are *computed*. To compute, we need a computer—a machine that converts premises into conclusions. But to have a machine that computes, the machine must itself have to obey some laws which in turn need to be computed in some yet another more generic computer, ad infinitum. The idea that nature is governed by mathematical laws is thus not as straightforward as it initially seems because the mathematical laws governing nature need to be computed, which requires a computer, which requires matter, which in turn must be governed by mathematical laws, and so on, ad infinitum. For the idea that nature is mathematical to be simple and intuitive, we require something that computes itself, but such a thing has not been found. Eugene Wigner called attention[1] to this "unreasonable effectiveness of mathematics in science" arguing that there is no reason why nature should be governed by mathematical laws, although we use mathematics in all physical theories.

Underlying this problem in physics is a subtle relation between *matter* and *law*. Matter is the stuff that makes up the universe and laws are entities that effect change. We equate these two things in practice, attributing laws to matter, but the truth is that they are different because laws require computation while matter does not. This distinction is very important from an informational standpoint because the information in matter can be treated as *data* while the information in laws needs to be viewed as formulae or program *instructions*. Formulae compute changes and we can think of information in terms of *functions* that compute via physical processes rather than using computers. This is possible if laws are viewed as a level of physical reality that accompanies objects. The critical shift needed to view quanta as computers is the ability to see the information in them both as concepts as well as programs. The same information needs to be dually-interpreted as data and instructions. This is possible if position and momentum in quantum theory can be interpreted as object-concepts and action-concepts, respectively. Matter and laws are not separate things as in current science, which postulates material particles and then laws governing them. Rather, laws should be viewed as programs present within matter.

Recall the discussion on Turing's theorem earlier and how the inter-conversion of data and program led to the Halting Problem. If inter-conversion is problematic in computing theory, how is it permissible in the case of physics? In quantum theory, object- and action-concepts are derived from two representations— position and momentum—of the same wavefunction. These representations are related by a Fourier transform and are therefore complementary descriptions of the same reality. Furthermore, the object- and action-concepts associated with the bits are derived from the context. The dual interpretation of a wavefunction as object- and action-concepts is not the same as that in Turing's case because of the context and because object- and action-concepts are complementary.

However, since quantum particles are in a stationary state, and cannot automatically transfer energy, they also cannot automatically compute. A quantum system has the *potential* to compute, quite like a program which can be executed in a computer. However, to execute the program, a quantum ensemble must interact with another ensemble. This interaction requires transfer of energy.

The laws of nature cannot be formulae that nature computes without us knowing where and how they are being computed. The laws instead must exist physically in nature as programs. The meaning in a momentum representation can be interpreted as a program. The problem of computation reduces to how information in matter can be programs. In a deterministic world an observer cannot program the universe to carry arbitrary information and programmatic instructions. But the quantum universe is different because the quantum universe is not in a definite state unless we put it into a definite state with programming. This means that the behavior of objects is not necessarily governed by the laws of current physics but can also include programs by which we can program matter. The only criterion for physical programs is that they be computable—i.e. that they will halt. This is equivalent to saying that the program solves a well-formed problem, that it constructs a cogent object through its transformations, or that it is meaningful.

Programs in Quantum Theory

The wavefunction is the division of reality into orthogonal parts, called eigenfunctions. This division corresponds to the everyday fact that, in describing nature, we first identify the distinct and mutually exclusive concepts in terms of which reality is described. We might designate these distinct concepts as the "axioms" or alphabets of our description. The Hamiltonian is the ability to divide reality into logically orthogonal constituents. In effect, the Hamiltonian specifies the language in which any other information is encoded.

The division of reality into orthogonal parts isn't static. The division can change when matter is redistributed in space. By this, the parts in terms of which nature is described change, the alphabets of description are altered and the descriptions using these alphabets are also changed. The Hamiltonian represents the division of reality into axioms or alphabets, and changes to the Hamiltonian imply a change to the axioms or alphabets. This change is represented by the Time Dependent Schrödinger Equation (TDSE henceforth).

$$H\psi(x,t) = i\hbar \frac{\partial \psi(x,t)}{\partial t}$$

$$H(x,t) = -\frac{\hbar^2}{2m}\frac{\partial^2}{\partial x^2} + V(x,t)$$

TDSE does not have an exact solution in a general case. To obtain exact solutions, we assume that eigenfunctions that satisfy the equation will be *separable* into space and time components.

$$\psi(x,t) = u(x)T(t)$$

Substituting this into the TDSE, we get the following.

$$-\frac{\hbar^2}{2m}\frac{\partial^2 u(x)}{\partial x^2}T(t) + V(x,t).u(x)T(t) = i\hbar \frac{\partial T(t)}{\partial t}u(x)$$

Dividing by $\psi(x,t)$ we get an equation that is separated in space and time variables, except for V's dependence on t. To obtain an exact solution, we assume that $V(x, t) \cong V(x)$, which leads us to two sides of the equation that change independently in x and t.

When two equations in different variables x and t are equal, they are separately equal to a constant. Let's suppose that constant is E.

$$-\frac{\hbar^2}{2m}\frac{\partial^2 u(x)}{\partial x^2}\frac{1}{u(x)} + V(x) = E$$

$$E = i\hbar\,\frac{\partial T(t)}{\partial t}\frac{1}{T(t)}$$

Some rearranging now leads us to separated equations as below.

$$Hu(x) = Eu(x)$$

$$i\hbar\,\frac{dT(t)}{dt} = T(t)$$

These two equations produce solutions of the following kind.

$$\psi_n(x,t) = u_n(x)\,e^{-\frac{iE_n t}{\hbar}}$$

In this solution, the eigenfunction is fixed, although its phase keeps changing. We can think of such a function as a vibrating membrane. The assumption that the Hamiltonian and its solutions are separated in space and time variables is called the Separation of Variables (SOV) hypothesis and using it, the TDSE can be broken into two separate equations, dealing in x and t variables individually. While the SOV hypothesis isn't incorrect, making this assumption in general is wrong. The solutions we get eventually are in part due to the assumption that we put into the equation in the first place. TDSE can also be solved without a SOV hypothesis in the following steps.

$$H\psi(x,t) = i\hbar\,\frac{\partial\psi(x,t)}{\partial t}$$

$$H\psi(x,t)\,.\,\partial t = i\hbar\,[\psi(x+\partial x, t+\partial t) - \psi(x,t)]$$

Transposing and rearranging we get the following.

$$i\hbar\,\psi(x + \partial x, t + \partial t) = H\psi\,(x, t)\,.\,\partial t + i\hbar\,\psi(x, t)$$

The above equation says that the new reality given by $\psi\,(x + \partial x, t + \partial t)$ is equal to the existing reality $\psi\,(x, t)$ plus a change caused by the Hamiltonian on the existing reality $-H\psi\,(x, t)$. ∂t. We can now think of H as a perturbation of the reality, or a *program* that modifies the input $\psi\,(x, t)$ to produce an output $\psi\,(x + \partial x, t + \partial t)$.

This gives us a new meaning for H. The operator that represented the ability to *divide* reality into orthogonal components in the time-independent case becomes an operator to *modify* reality in the time-dependent case. The energy in H that distinguishes matter into distinct objects can also be treated as the ability to modify the nature of reality. The former represents knowledge of matter and the latter represents programs within matter. The former will divide nature into atomic concepts and the latter into atomic changes. Of course, as we saw above, this change is only the *potentiality* in a system and cannot be effected without an external change in field. This is like saying that the gun can be fired if someone presses the trigger. The gun itself does not dictate that the trigger will be pressed. The gun only defines the behavior *if* the trigger is pressed.

5

Advanced Quantum Topics

God knows I am no friend of probability theory, I have hated
it from the first moment when our dear friend Max Born
gave it birth. For it could be seen how easy and simple it
made everything, in principle, everything ironed and the
true problems concealed. Everybody must jump on the
bandwagon. And actually not a year passed before it became
an official credo, and it still is.

—*Erwin Schrödinger*

Quasi-Particles

The laws of motion in classical or quantum physics can generally
be solved exactly for single or small numbers of particles. When the
number of objects increases, exact solutions to multi-body problems
become very difficult. To solve these equations, we must use new
mathematical techniques, which convert a set of equations about the
interaction between various particles into equations that describe the
motion of independent but hypothetical particles. A two-body interaction problem in classical physics can for instance be broken into the
independent motion of two fictitious particles—the center of mass
motion and the displacement vector motion. Once separated, the
equations can be easily solved. This mathematical trick poses important questions about how we *observe* the world as measurements versus how we *describe* it in terms of concepts. The original equations
of motion pertain to measurement while the transformed equations
pertain to a conceptual description, and the particles underlying

these descriptions are quite different. We now have a clear difference between sensual and conceptual knowledge.

A similar mechanism is also used in quantum theory to solve N-body problems. Instead of describing individual quanta, we can describe the ensemble as a collection of fictitious particles. Such particles are called *quasi-particles* because (a) they are fictitious and (b) short-lived. Depending on the type of ensemble being described, many quasi-particles have been identified—phonons, chargons, magnons, spinons, plasmons, etc. Unlike classical physics where the mathematically transformed equations did not yield any new observables, in quantum theory most of these fictitious particles have been empir-ically observed. This raises important philosophical questions about which of the descriptions of nature is primary: the one that deals with electrons and photons or the one that describes nature in terms of quasi-particles? Within quantum theory we can confidently say both quasi-particles and particles are real because they represent an orthogonal basis in the wavefunction, although a similar claim in classical physical would have been impossible.

The notion of an individual object in quantum physics depends on how energy is distributed, and there are many ways to cut up the ensemble into particles by redistributing energy. Schrödinger's equation provides an orthogonal basis for *logically* cutting up the ensemble into distinct particles, like we would reduce a language into alphabets or mathematics into axioms. This flexibility also allows us to divide the world at several different *levels* of abstraction. Theories in physics, for instance, can use one set of concepts, in chemistry another set, biology a different set, and so on. The concepts of biology may be composed of concepts in chemistry and the concepts in chemistry may be composed of concepts from physics. However, this does not make the chemical or biological concepts any less real. There are many orthogonal ways of cutting up the ensemble and they are equally real. Quasi-particles are a description of the ensemble at a higher level of abstraction than sub-atomic particles; these are quantum objects, but they are also 'macroscopic' particles.

We can describe nature in terms of different axiomatic sets, as each eigenfunction represents a single conceptual symbol. A given axiom set is one way of cutting up the world into fundamental objects, and the ability to change these axiomatic bases is a new type of conceptual relativity, that we did not have in classical physics. For instance, even after the N-body problem is solved in classical physics, the resulting objects are still conceptually the same—particles. In quantum theory, since objects denote different meanings, changing the eigenfunction basis also changes the *types* of these particles.

Schrödinger's equation should be viewed as a method by which to reduce an ensemble into its orthonormal parts. The world is comprised of epistemically and not ontologically real components. This shows the primacy of concepts over things in quantum theory. In classical physics, things are primary and fictitious particles are real in a mathematical theory but unreal in fact. That is, the identity of objects is given by *a priori* real things and not by concepts. In quantum theory, the identity of the particles is not given by *a priori* real things but by the theory. Any physical system can be divided into orthogonal parts like the alphabets of the descriptive language *in terms of* which we must describe that system. Since these components can change, the description then would be given in terms of different concepts. This corresponds to transforming a set of concepts into another, and describing nature in terms of different concepts, none of which seem more fundamental than the other.

Quasi-particles represent the fact that nature is not *a priori* cut up into electrons or quarks as fundamental particles. Rather, as ensembles of quantum objects are collected, what is fundamental in nature can be differently based on ensemble constitution and depends on the *choices* of a basis. This fundamental reality is conceptually and empirically real and therefore we must suppose that nature *is* comprised of that reality at the point of observation.

Quasi-particles also tell us that an N-body system can be viewed as a smaller set of particles. That fact can help us relate quantum theory to the macroscopic world. So far, quantum theory is regarded as a theory of the sub-atomic particles. The everyday world has large ensembles of particles and their quantum behaviors are not visible. We designate these large ensembles with everyday concepts—tables, chairs, cars, etc. A semantic view of quantum theory combined with the notion of quasi-particles tells us that macroscopic objects can also be viewed as *quantum* rather than *classical* objects. These objects will represent macroscopic rather than microscopic symbols. The eigen-functions of macroscopic quantum objects will obviously be larger, but they too will have position and momenta that denote semantic properties rather than physical properties. An atomic object can be a symbol of a macroscopic object, and a small macroscopic object a symbol of much larger macroscopic objects.

This is important because it attacks the classical-quantum divide. Quantum theory does not reduce to classical theory, because there are quantum effects seen even at the macroscopic level. In fact, the use of orthonormal eigenfunctions at macroscopic level will divide reality into *logically* distinct *types* of matter. Given the nature of interactions within the ensemble, the logically distinct components are defined by the rules of quantum theory.

By applying quantum theory to the macroscopic world, we can derive some new conclusions. First, a large ensemble must divide into orthonormal components; the universe as a closed system is com-prised of *logically* distinct components, which must be understood via different concepts. We should not think of the universe as a uni-formly spread matter, but as comprised of logically distinct ensem-bles. Second, the identity of these distinct ensembles is again defined in relation to other ensembles, which are combined to form a larger ensemble. This constructs a hierarchy of ensembles and the whole is reducible to parts, but the parts are not independent of each other. As a system grows in complexity, it must reorganize, simplify and reduce into a new orthonormal basis of types. This generic principle can be applied equally well to microscopic and macroscopic worlds, and quantum theory describes both.

Our ability to describe nature in many ways leads to some wor-risome questions. If the same reality can be differently described at microscopic and macroscopic levels, then are these ways equivalent such that no theory is more fundamental than others? If nature is described in many equivalent ways, then aren't they all equally funda-mental? Why is one view more basic than others?

Our ability to use different concepts to describe nature does not preclude the existence of fundamental theories. A fundamental the-ory is that which cannot be further reduced into smaller parts. The quantum of action represents a limit to the divisibility of nature into

parts. The smallest particles in nature are therefore the most fundamental as well. While nature can be described using other concepts,
these atomic particles remain the most fundamental descriptions.
The semantic view allows us to describe nature in both atomic and
macroscopic concepts, although the atomic concepts remain more
fundamental than the macroscopic ones. These atomic concepts will
represent the most fundamental ideas in nature.

Schrödinger's Cat Paradox

Schrödinger's paradox states that if a cat should be imprisoned in a
chamber with a poison vial that could break anytime depending upon
a quantum radioactive decay event, then if we were to observe the cat
at any random point in time (before or after the decay), quantum theory would be unable to predict the objective state the cat is in—dead
or alive. Accordingly, the cat is in a superposed state of being dead or
alive and different observations must sometimes reveal a living cat
and sometimes a dead cat. That is not just unintuitive but also contradicts observed experience. Once the cat is dead, we should never
observe a living cat. There must be a real physical transaction that
takes the life out of the cat at some time.

Why is the cat thought experiment such a problem? It is a problem only if we suppose that the wavefunction is in a superposed state
because in that state we will sometimes see a dead cat and at other
times a living cat. That prevents us from talking about the real state of
the cat, because the cat is always in a superposed state.

In quantum theory, the wavefunction must change in time to
describe the cat experiment. The eigenfunctions of the system prior to
decay will include the possibilities of detecting {living cat, unbroken
poison vial, un-decayed atom}. After the decay, the eigenfunctions for
the system will change into the possibilities of finding {dead cat, broken poison vial, decayed atom}. Two things must be kept in mind here.
First, the cat can't be described in isolation of the rest of the experiment, and the wavefunction can't be formed by superposing a dead cat
with a living cat. The wavefunction must include the poison vial, the
radioactive decay setup, etc. This means that when the cat transitions

from alive to dead, the poison vial transitions from unbroken to broken, the radioactive atom transitions from not-decayed to decayed. Second, the transition is *caused* by an event of radioactive decay which can't be predicted in advance today. While that event is intrinsic to the whole setup, it is a change in time nevertheless. To solve the cat paradox, we must find a way in which to conceive of the radioactive decay as a cause that changes the whole wavefunction in a time-dependent manner.

The mysterious element in this whole setup is not whether the cat is alive or dead, but whether the radioactive atom has decayed or not. Because, if we substitute the radioactive decay with an electric discharge caused by an experimenter outside the system, which takes the cat from alive to dead, there would be no mystery. We would now attribute the cat's death to the press of a button by the experimenter, which would then be explained by other changes in the experimenter's mind and body, all the way to his free choice.

Whether the cat is killed by a human choice or by a radioactive decay, there is a change in the system unexplained by present quantum theory. What makes both unintuitive at this point is because they are *discrete*—we can't pin the changes that incrementally take the previous state into the next state, through infinitesimal steps. To account for this, we need a theory of discrete changes. With a theory of discrete changes that can occur spontaneously (and not triggered by external agency), we would not need to think of the cat in the superposed state at all. We would then be comfortable in thinking of a living cat that dies because of a discrete change. The need for quantum theory and its interpretations is thus to produce a suitable account for spontaneous discrete changes. This seems challenging because our notion of causality involves actions by external agencies and in the case of radioactivity there is no external agent.

The cat paradox is thus hard because it introduces a radioactive decay, not because it is any harder to understand as compared to other quantum phenomena. The problem of spontaneous occurrences is however not unique to radioactivity, and it can be seen even in the 2-slit experiment where quanta arrive spontaneously at different detectors. Which detector will fire next cannot be predicted by incremental steps. Due to this, we cannot describe the *sequence* of symbols observed in the 2-slit experiment. To avoid probability, we need to identify something objective that corresponds to the sequence being observed. The cat paradox, radioactivity and the 2-slit experiment are all probabilistic for the same reason—even with a semantic view of nature, we still haven't explained the time ordering of events in nature. In the cat paradox, we combine probability with superposition and arrive at the unintuitive idea that the cat is both dead and alive. We can resolve this problem by removing superposition and adding an event ordering. Now, the cat dies due to radioactive decay, which changes the total wavefunction in which the cat's eigenfunction goes from living to dead. The question is: How would we add event ordering to quanta?

Event Ordering

Event ordering is a widespread phenomenon in quantum theory. Aside from radioactivity, it also exists in the order of quantum arrival in the 2-slit experiment. If quantum theory could understand and predict this event order, then it will not be a probabilistic theory. I have earlier argued that event order can be understood when quanta are treated as symbols. The order now represents a meaningful proposition or a meaningful program. The quanta in this proposition or program are alphabets or instructions. Their order represents a *procedure* by which a complex program or description can be constructed by sequencing elementary semantic symbols.

In ordinary language or programming, a proposition or program represents information encoded by an author or programmer. While this information can be understood, there is currently no way to predict it. Indeed, Shannon's Information theory argues that if the information could be predicted by the receiver, then there would be no need for a sender to send the information because the receiver could predict it anyway. Shannon argued that only the bits that cannot be predicted must be transmitted because they constitute the 'uncertainty' in the receiver's state. The theory of information and classical physics therefore lie at opposite ends in terms of predictability. In classical physics, given an initial state any subsequent state can be predicted. In information theory, every unique event must be unpredictable because it does not follow from the previous event. I will, however, now argue that even information transactions can be predicted although this requires a different view of *time* than is held in current science. To understand this new view of time, let's examine how time is seen in everyday life.

While time in science is *linear*, time in everyday life is *cyclic*. Common experiences of cyclic time include the passing of minutes, hours, days, weeks, months and years. Biological systems work in a cyclic manner and cyclic behavior is seen in the rise and fall of ideologies, societies and civilizations. If you know that the clock's hour hand points at 10 o'clock right now, it will point at 11 exactly 60 minutes from now. If you know today is Thursday, it will be Friday tomorrow. If you know it is morning right now, then you can also predict afternoon, evening and night. This ability to predict, however, rests on an understanding of time's cycles and where we currently 'sit' in that cycle. If you know the location in the time cycle then you can predict the future events; otherwise, you cannot.

We saw earlier how semantic notions about space arise in a closed space. For instance, the locations in a house are defined as kitchen, study, living room, bathroom, bedroom, etc. In a similar fashion, time is defined semantically when time is cyclic. Hours in a day, days in a week, weeks in a month, months in a year and years in millennia are all semantic ways of describing time. Biological circadian rhythms are semantic ways of describing chemical activity in terms of breathing, digesting, eliminating, circulating, etc. Human activity that follows cycles is also described semantically. There is, hence, a close relationship between the semantic order in events and the cyclic nature of time. I will therefore hypothesize that the order in information can be predicted by a theory if the time in that theory is viewed cyclically. The order of semantic events will then be the order of events as predicted by a cyclic order in time.

We must, however, note that the universe can be described by several nested levels of cyclic time. This is like how space can be described in terms of nesting. For instance, my house is within a city area, which is in the city, which is in the state, which is in the country, which is on a planet, which is inside the solar system, etc. Similarly, the moment now is embedded in the minute, which is in the hour, which is in the day, week, month, year and so forth. To understand the time cycle we need to know both the level of nesting as well as the current semantic temporal state in each level of time nesting. For instance, I cannot completely tell the current time simply by the seconds on the clock hand. To completely describe the current moment, I also need to specify the minute, hour, day, month and year. Once we know the time nesting hierarchy and the state of time in each of these nested cycles, we know the current time.

Currently, time is either described in relation to the start of the universe or in relation to some arbitrary point in history. Physical theories can, for instance, speak about the origin of the universe a few billion years ago. Computer clocks are, on the other hand, synchronized with the start of January 1, 1900 which is an arbitrary reference for time. Both the absolute time of start in a linear time and an arbitrary time of start in a cyclic time are not useful for making semantic predictions. To make cyclic predictions, we need an absolute time of start in a cyclic time and understand the manner in which time is further divided and nested within higher cycles.

Such a cyclic time, together with a closed space, can be tied to semantic events. For instance, a detector at a specific location in a closed space will denote a concept and the time of that detector detecting a quantum will represent an event. Current quantum theory can predict the probabilities of locations but not the order in them. In current quantum theory, closed space has been incorporated through the notion of an ensemble, but cyclic time has not yet been incorporated. The causal predictions of quantum theory using linear time are therefore arbitrary. If locations in a closed space are understood semantically, then the order in these events can also be explained provided time is viewed cyclically. Quantum predictions will now be as deterministic as classical physics. Just as knowing the initial state allows us to predict subsequent states in classical physics, knowing the location and time of start for an experiment will allow us to predict the succession of events in quantum theory.

If quantum theory is a theory of meanings, then its description cannot be completed only by treating space semantically. Rather, both space and time need to be revised. This means that space should be divided into smaller and smaller regions, which are nested within larger regions. Similarly, time should be divided into smaller and smaller durations which are nested within larger durations. Quantum theory stipulates that there is a limit to how small a region of space and time can get. This smallest region of space-time represents the most atomic 'event' in the universe. However, since this event is defined by dividing a much larger domain of space-time into smaller parts, each of the space-time domains in the hierarchy can be given semantic properties. It is therefore incorrect to assume that the atom is semantic, although the macroscopic world is classical. The atom is a semantic event which is structured through space-time nested hierarchies to construct larger and larger phenomena. All these phenomena must be described semantically.

In the case of the double-slit experiment, the order of events detected during the measurement will depend upon the time at which the experiment is started. The ensemble is like a book or program which was created by ordering the symbols in a certain order. The order embedded in the ensemble is a consequence of the semantic order in time at which the system was prepared. But the order detected during measurement is a function of the time at which the measurement is performed. In effect, the measurement may not read the book or execute the program from the 'beginning'. The measurement may as well read the book or execute the program from somewhere in the middle. The point at which we start reading the book or executing the program depends on the time when the measurement is started. This means that the observer of a semantic system will get the same *knowledge* (if he or she decodes the system completely) but may not get the same *experience* because the order in the events is altered based on the time of observation. Of course, it is possible to read the books in just the way they were written, but to achieve the similarity between the author's experience of writing and the observer's experience of reading we would have to also align the time cycles during state preparation and observation.

When this view is applied to radioactivity, the apparently random decay of radioactive atoms is contingent upon the time cycle. The randomness of radioactivity is therefore apparent. We are observing what was put into the world through state preparation, but the event of observation is contingent upon the time cycle of observation. When a cat dies due to a radioactive decay, the time of death cannot be predicted by current quantum theory. But the time of death can be predicted by a semantic quantum theory because the order of events (which include the radioactive decay, the breaking of the poison vial and the cat's death) will follow time's cycle.

SI gives us some insights into *why* the universe evolves in time: while information can be stored in space through spatial locations, it can't be created or communicated in that way. To create and communicate, we must *serialize* information. This problem has been studied in computing and communication theory. The problem of communication entails that information must be serialized or converted into a *description* or a step-by-step *procedure*. Serialization implies that the state preparation is done step-by-step such that the constructed object can be deconstructed into steps, transmitted as steps, and then constructed back into the original in steps. The steps in the creation of information may not be synchronized with the steps in observation and what we observe may not exactly be how it was previously encoded. But the *knowledge* communicated through such out-of-step encoding and decoding will remain unchanged.

The Problem of Dimensionality

In physics, space has three dimensions and time has one dimension. There are exotic theories like string theory that extend to as many as ten dimensions, but by and large, most of physics assumes a total of four dimensions: length, width, height and time. Why does space have only three dimensions and time only one dimension? This question is very old and has got no answers from fundamental theories. We just take this as a fact about the present universe.

The relation between space-time and information can unlock this question in a completely new way. If nature is symbolic, then a dimension in space or time must be useful in *representing* some aspect of information. The question about the number of dimensions required thus boils down to: How many dimensions are needed to completely represent all sorts of sensation and action concepts in space and time?

Note that the ability to represent concepts in space-time is necessary for us to represent information in the real world and to *communicate* it to others. If something can be represented through a changing material object, then it can be represented in space and time because that material object can be *described* by another object by reinterpreting space and time informationally. This leads us to the conclusion that space and time are necessary and sufficient to represent all sorts of information, and the total number of dimensions needed in a physical theory should be sufficient enough to represent informational variety and no more.

This means that we need a certain number of dimensions to represent all sorts of concepts, but also that these dimensions must be given an informational interpretation. That is, the dimensions should not only be denoted as physical arrows but be described as some sort of informational property useful to represent concepts. When material objects are used to represent information, then the ordinary space-time is also a semantic space-time in which the evolution of objects represents the evolution of ideas. For ordinary space-time to be used to represent ideas, its properties must be necessary and sufficient to describe all type of concepts. This is a novel approach to the question of dimensionality because it brings in the necessary and sufficient condition not from the perspective of physically observed facts about the universe but from the angle of what sorts of concepts need to be expressed through space-time. The dimensionality of physical space-time is now subject to the extent of the conceptual universe, provided, of course, that we can physically express and communicate everything that we can think.

Dimensions in geometry are used in two ways—(a) as an ordered set or points and (b) as representing magnitudes or distance. If the universe is comprised of discrete events, which can be distinguished by some scheme, then all events can in principle be counted, labeled by natural numbers, and placed on a single dimension in an ascending order from left to right. In short, to order all the events in the universe, we need only one dimension provided the universe consists of discrete and countable events. Quantum theory tells us that the universe is discrete, which means that we need a single dimension to order all events. But this single dimension belies multiple orthogonal dimensions which must be used to distinguish concepts from one another *before* they can be ordered and counted. The single dimensional ordering is a *result* of distinguishing along multiple dimensions by which we would reduce those dimensions to a single dimension. Any finite number of dimensions can be reduced to a single dimension if each of those dimensions comprises discrete and countable number of points[1]. But that single dimension on which we order events would be distinct from the dimensions on which we distinguish those events.

When I speak of dimensions needed to represent information, I mean the dimensions needed to *distinguish* events rather than to *order* them. If the universe consists of discrete events, only one dimension is needed for ordering events although multiple dimensions are still required for distinguishing those events. I don't intend to answer the question of dimensionality here because answering this question requires an exhaustive analysis of object and action concepts—and that will take us away from the primary purpose of interpreting quantum theory. My goal is to show that this question can be tackled and answered in a new way using a semantic approach to quantum theory. Whether this analysis confirms that there must be four dimensions, I will leave it to a later effort.

Coherence and Decoherence

A common problem in interpreting quantum theory is how a quantum system interacts with other quantum systems. The origin of the problem is that once two quantum systems have interacted we are forced to think of them as a single system and not as two separate systems. This is sometimes called entanglement and at other times simply a quantum ensemble. What causes entanglement, and why can't we treat the interacting quantum systems separately?

Assume we have a system $S1$ given by the Hamiltonian $H1$. It has orthogonal eigenfunctions $\{\psi i\}$. Now we take another system $S2$ with a Hamiltonian $H2$ which has orthogonal eigenfunctions $\{\varphi i\}$. If $S1$ and $S2$ were made to interact, there are two ways in which we can conceive of this interaction. First, the combined eigenfunction of the system can be a product of the individual eigenfunctions $\chi = \Psi \otimes \varphi$.

Eigenfunctions of this type represent separable systems. Second, we can take the individual Hamiltonians of the two systems (H1 and H2) and arrive at a new joint Hamiltonian HN and then use it compute the eigenfunctions of the combined system. Since the two systems are interacting, their combined Hamiltonian must have an interaction component which is absent from the Hamiltonians of the individual systems. Correspondingly, when we compute the eigenfunctions using HN they will be different from the *pure* eigenfunctions earlier computed using H1 and H2. The combined system is now entangled because we cannot describe S1 and S2 independently. They must be described as one system SN whose states are given by the joint Hamiltonian HN and not by separate Hamiltonians H1 or H2.

If the two interacting systems were loosely coupled to begin with, and their interaction was weak, then the eigenfunctions in the joint wavefunction would look like the separately orthogonal states of the two independent systems. We could then imagine separating these two systems by drawing them apart which would remove the interaction terms in the Hamiltonian and the two systems would be disentangled. But it may not be always possible to disentangle quantum systems in this way. Particles with opposite spins, for instance, cannot be decoupled in this manner. They remain entangled even when they are spatially separated, and the problem of non-locality arises due to this entanglement, as seen previously.

There is another interesting possibility called decoherence, where an observer and observed systems are not merged into a single larger system but remain distinct systems although the states in the two systems are *aligned* one-to-one. This corresponds to the case where we try to know the world according to a set of concepts {C_i} and the world is described by those very concepts. In this case, a measured system has a set of eigenfunctions that correspond one-to-one with the eigenfunctions in the measuring system. In achieving this alignment between the states of the measuring and measured systems, the states in *both* systems must be changed. This is somewhat counterintuitive from the standpoint of classical physics. In classical physics, we assume that the measurement reveals the nature of reality as it existed prior to being measured. Thus, during measurement, the state in the measuring system must change although the state in the measured system must remain unchanged. Quantum decoherence violates this principle. Now, during interaction, the state of the measured system is also changed to align with the states in the measuring system. In other words, the world is known in the manner in which we try to know it and it reveals not properties that might have existed in the system prior to being measured but in a manner similar to what is being measured.

A simple way to illustrate this idea is to consider what happens when we look at the world through colored glasses. If the glasses are tinted red, then we will see the world as red. However, we do not interpret this to mean that the world is indeed red. We only suppose that our perception is red although the world is not red. In quantum decoherence, if you see the world red, then the world is indeed red. The manner in which we perceive the world—e.g., by viewing it through red glasses—changes the way the world exists. Of course, the world also changes our methods of perception, and if the world was green, it would—during decoherence—change the color of the glasses. Decoherence is thus a phenomenon in which our glasses become somewhat green and the world becomes a little red. The two systems align their states such that if we look at one of the systems we can know about the other, since their states are aligned.

Decoherence is the conversion of a quantum system into a classical system and the way this happens is as follows. Assume that there is a measured system whose eigenfunctions are represented as $\{\varphi i\}$ and a measuring system whose states are represented as $\{ei\}$. When the measured and measuring systems interact, their eigenfunctions are not divided into an orthogonal set governed by a combined Hamiltonian. Rather, the states in the two systems evolve slightly to become $\{\varphi'i\}$ and $\{e'i\}$ such that these two sets of eigenfunctions are one-to-one correspondent. The wavefunction χ of the total system can now be written as $\varphi'1.e'1 + \varphi'2.e'2 + \ldots + \varphi'N.e'N$. If a measurement was performed on this total system by an observer, the result would be one of the $\{\varphi'i\}$ observed. The reason this represents a classical system is that if we computed the transition probabilities of the measured system given by the state φ going to a new quantum state Ψ then this would not have any of the cross-terms, which are the basic hallmark of quantum interference.

The reason for this observed behavior is that the concepts represented by the measuring system (the environment) do not change when the measured system goes from φ to Ψ. The environment's orthogonal basis is the same and the eigenfunctions in the measured system must align with this conceptual basis before and after the change. It follows that when φ changes to Ψ it cannot evolve into a new conceptual basis because the conceptual content is already fixed by the environment. What can change are relative proportions of $\{e'i\}$. When the conceptual basis in a system is fixed, then effectively the system appears to behave like a classical system. The measured system is not actually a classical system because it has conceptual content, but it appears to exhibit classical behavior because it is described using a pre-defined set of concepts.

In effect, the *vocabulary* in the measured system changes to suit the vocabulary in the measuring system. What is said using these vocabularies (event sequences) is still outside present quantum theory. Thus, the view that decoherence makes a quantum system classical is only apparent and it arises because we fix the concepts by which we will describe the measured system, which forces the conceptual content in the measured system to be reorganized such as to align with the manner it is being measured and described.

Through such reordering of content, it is inevitable that the $\{\varphi'i\}$ are actually *mixed* states constructed out of combining $\{\varphi i\}$. This is yet another reason to think that the system becomes classical as it now appears that we are observing the superposed state rather than the pure states. One key problem that physicists have worried about is why a quantum system must always be observed in one of the pure states rather than the entire superposed state. That is, can the wavefunction as a whole correspond to something classical? While the total system is described by the wavefunction $\Sigma ai.\varphi i$ the whole wavefunction apparently represents an ensemble of particles which can be known classically but not quantum mechanically. The observation of the mixed state would correspond to the macroscopic system while the observation of pure states corresponds to the individual quanta. If we observe the system in a mixed rather than pure state, then we have transitioned from quantum to classical.

This picture of quantum to classical transition is mistaken as the individual states represent concepts which need to be combined in a specific manner using event ordering to arrive at the total meaning denoted by the wavefunction. Measuring the superposed state corresponds to measuring symbols without knowing how they are *ordered* in ensembles to form meaningful descriptions. That measurement can tell us what kinds of symbols are being used but it is still not indicative of the actual meaning denoted by them.

Decoherence essentially makes the world align to the manner in which it is being observed. If the measuring system was designed to detect the world according to a prior fixed set of concepts, only those concepts would be observed and the conceptual content in the measured system would be modified to suit the measurement apparatus. This is a new type of coupling between measuring and measured systems and it stands upside down the classical view where the world is objective, and we know the world as it exists prior to being known. Decoherence basically tells us that the world may be known according to the way in which we try to know it. The measured system somehow learns about the way in which the measuring system is trying to describe it and then adapts to it.

As the saying goes, if we only have a hammer the world appears like nails to us, and this type of coupling between the knower and the known implies that even if the world was not nails, then it would be observed in terms of nails when the measurement is performed. The knower and known in this case are distinguished by the observer-observed distinction that Niels Bohr made in the early quantum theory days but equated it with the classical-quantum 'boundary.' In the case of decoherence, there is a different kind of quantum boundary that employs a uniquely quantum theoretic semantic distinction. This has profound implications for science because if the world can be known in the way we want to know it, then science is not just about an objective discovery of reality but also an ethical choice about the types of concepts in terms of which we choose to describe nature. Of course, it would be wrong to assume that the observer imposes their thinking on nature. The reverse is also true. The combined wavefunction of the observer-observed system must evolve into a shared basis of concepts that are similar across both. The observer-observed boundary therefore determines a shared conceptual basis in terms of which both systems will exchange information. This basis is orthogonal, like any quantum system.

This view of entanglement can explain how an observer can know the mental states of another observer even though the two observers cannot experience each other's mental states. Each observer's brain in such a case represents the *contemplation* of some facts outside the brain, but not a direct sensual involvement. That is, I think that I know what another person is thinking, although I cannot experience that thinking. Similarly, I may be able to shape another person's thinking without my senses and mind directly acting on someone's senses and mind. Distinct ensembles in the case of decoherence maintain distinct identities and don't collapse into a single ensemble. And yet, these ensembles exchange information, unlike the classical notion of communication supposes. In classical communication, distinct objects exchange information by emitting and absorbing data, but without forming a single, larger ensemble. The emitted information travels with a finite speed and arrives unchanged. In quantum communication, objects communicate by becoming part of a single ensemble, and they are shaped by other objects in the ensemble. The classical notion of communication therefore does not apply within a quantum ensemble. A symbol that moves from location P to Q in space will transform such that what the receiver obtains is different from what the sender sent. But so far in SI we have discussed the behavior of a quantum object within an ensemble, whereas decoherence involves a communication between two ensembles. To understand this within the context of quantum theory, we must postulate the existence of ensembles that contain other ensembles, much like sets can contain other sets.

The relations between ensembles are like the relation between a picture and reality. The reality and its picture are correlated one-to-one, but the picture is not reality. Similarly, a proposition that describes facts is correlated to the facts, but it is different from the facts. A picture and the reality it depicts are therefore in the same ensemble, but the *relation* between them is not that of a *logical distinction*. It is rather an *intentional* relation where one object refers to and describes another object, without being that object.

Decoherence needs a widening of the notion of ensembles from systems that contain logically distinct objects to systems that can

be intentionally related. Intentionality is not just a human phenomenon where we create pictures and books about reality but exists
even within atomic systems. Parts of an intentional system communicate by transforming each other's states and getting entangled in
a way that one system describes the states of the other system. This
entanglement is different from that within an ensemble where objects
are epistemically distinguished. Intentional entanglement involves a
complementary intentional relation wherein two systems are distinguished as a knower and known, although the knower refers to the
known and the known refers to the knower.

The interacting ensembles are still quantum entities, and not classical objects. The nature of the quantum system is, however, different
from the one that involves just one ensemble. The knower and known
must now transform each other without emitting a traveling particle. This transform is action at a distance and behaves non-locally,
although it takes a finite amount of time (depending on the semantic difference between knower and known), which can look like information is traveling from the source to the destination. Decoherence
represents a new type of quantum system in which objects are being
distinguished intentionally rather than logically.

Ensembles and Pictures

The states of quantum particles in a picture will be correlated with the
states of particles in the reality that the picture depicts. But, if we don't
know which picture is correlated with which reality, then how do we
acquire knowledge of reality by looking at the picture? For instance,
you can look at a picture of Mona Lisa, but if you don't know that this
is the picture of the real Mona Lisa, you cannot obtain information
about the real Mona Lisa. To know that some picture describes Mona
Lisa there must be something in the picture that relates to the real
person that it aims to describe. Without such a relation, the picture is
yet another object. But with the relation, the picture becomes knowledge about some reality. Obviously, if the object that is pointed to by a
picture does not exist anymore—as in the case of the picture of Mona
Lisa—then how do you point?

This problem is solved in ordinary language by *naming*. We *name* the picture of Mona Lisa by calling it 'Mona Lisa.' The picture therefore does not actually *point* to the real person depicted in the picture. The picture however calls out the person through a name.

If quantum ensembles must depict pictures of reality, then they must encode *naming* in addition to meaning. Without a name, a picture can provide information, but we cannot know what this information refers to. And if we cannot know what the information refers to, then we also cannot know whether the information is *true*. Naming therefore enters language because, by referencing objects in propositions, the proposition intends other objects. This is a good way for us to learn about the object, but it is also necessary for us to confirm the truth of the intending proposition. The existence of a quantum ensemble only provides us with some meaning, but we cannot know if that meaning is true unless we know the naming.

We saw above how quantum locations depict concepts and temporal order amongst these locations depicts propositions. For a quantum ensemble to refer to other objects, additional quantum properties are needed, which can be interpreted as the *name*. We earlier discussed how an object can be completely distinguished by its location in space. The location in space can therefore be used as *name* to identify that object. If this location is given a number, then than number becomes the name by which the object can be called. For an object to refer to another object, this number (which identifies its location in space) must be encoded in another object through a new quantum property different from space and time.

This naming requires us to postulate two additional 'spaces' beyond the conceptual space. Recall the earlier discussion of three kinds of spaces—universals, individuals, and relations. The symbols are combinations of universals and individuals, but we generally neglect their individuality while speaking about their meaning denoted by the universal. However, while naming objects, we must invoke this individuality. Furthermore, to *refer* to such individuals—e.g. as knower and known—we also need to invoke a relation between these individuals. The knower and known prior to the relation are distinct, but they are related through knowledge, which is a relation between the two. Therefore, knowledge requires the combination of universals, individuals, and relations. The individuals are required for the knowledge to pertain to an objective reality different from the knower. The universals are needed for this objective knowledge to be symbolic and represent meanings. And the relation is necessary for a selective apprehension by a knower of some known (as the knower may not know everything).

As we discussed earlier, each of these spaces are possibilities. They are real in the sense that they exist, but they cannot be known until they combine. Thus, it is possible to speak about a concept that exists but may not be known to anyone. Similarly, a concept may not be realized into an individual instance of the concept. And even if it is realized, there may be no knower who has a relation to this individual for knowing. Thus, the combination of the three spaces creates the phenomena, which we call 'knowledge', but the reality underlying these phenomena exists even when the phenomena don't. By drawing this distinction between reality and its observation, we can solve the problem of naming. The problem comprises of two parts. The first part is identifying an individual by a name, and the second part is having a reference to that name through a relation. The former pertains to individuality, the latter to a relation. The meaning signified by that individual pertains to the universal.

The pointer from the knower to the known (or vice versa) is not a pointer in the classical sense of a vector. Such vectors are unidirectional, and they entail that the identity of the knower or known is unchanged in the process of knowing. For example, if the vector points from the known to the knower, the knower can change in the process of knowing, but the known will remain unchanged. This, as we have seen in the case of decoherence, is incorrect. The process of knowing alters both the knower and the known, and this is possible if we conceive of knowing as a bidirectional relation.

A picture of some object has the universal and the individual, and the relation to the object is established by the artist who represents the object into a picture. The picture is in some sense a 'knower' of the object. We can compare the picture to a measuring instrument that represents the reality being measured. By this process, the universal in the object is embedded in the picture, although they are separate individuals. So, in one sense, the picture names the object being represented, and in another sense, it is related to the represented object.

Classically, the picture and the object it represents are separate individuals. But, as quantum objects, one object is related to another as a symbol. It now becomes possible to describe the Himalayas without the picture itself being the Himalayas. This would be impossible in classical physics where an object's state cannot be 'correlated' to the state of any other particle and one state cannot be viewed as a description of another state. Quantum decoherence allows such correlations and it becomes possible to view the state of a particle as a symbol of another particle.

The Speed of Light

Einstein treated the constant speed of light as a law of nature in his relativity theory based on the Michelson-Morley experiment. But the idea that light has a speed of motion predated the development of quantum theory. If light is a quantum object, then its speed must also have some quantum explanation. Such an explanation doesn't yet exist. Indeed, the theory does not permit motion and speed in the way these existed in classical physics. In quantum theory, unless an object's energy is changing it remains in a stationary state[2]. The motion indicated by the speed of light across galactic distances therefore is somewhat of a mystery in the theory that describes ensembles of particles. I will try to demystify that mystery here.

The paradox of a constant speed of light is that the *time* taken to traverse from one *object* to another *object* is constant. Traditionally, speed is defined based on the time taken from one *point* to another. If an object moves from one point to another, then the time taken is computed based on the *points* and not the *objects*. This is illustrated in Figure-13. Assume that a photon P travels from a source S to a destination D and that both objects are not moving. In this case, assume that it takes time *t* for the photon to reach D from S. Now, classically speaking, it would seem that if D was moving towards P it would take P a lesser time to reach D. If instead D was moving away from P, it should take P a longer time to reach D. However, experiments show that even if D is moving, it still takes the same time for the photon to reach D. This is called the constant speed of light.

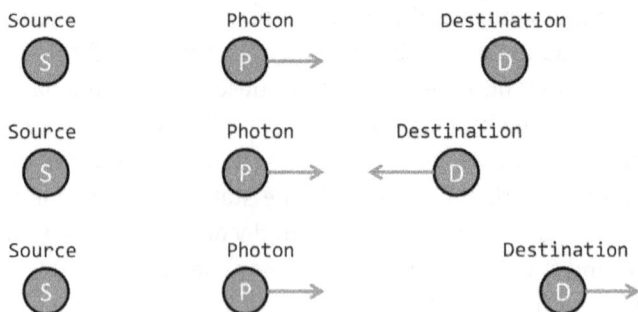

Figure-13 Light Takes a Fixed Time to Reach an Object

The constant speed of light seems bizarre only if we think that the light travels *in* space and therefore reaches a *point* in space rather than an *object* in space. This problem would not arise if light went from *object* to *object* rather than from *point* to *point*. In this case, the motion of the object would be immaterial for the photon, since the photon is headed towards the object and not to a point. But to hold such a view, we would have to think that the photon is 'headed' to an object—i.e. it has a predetermined destination. This creates a problem, namely, that the photon appears to have a goal. How can there be goals in quantum particles? Quantum particles don't need to have a goal if we think of a quantum particle as a *transformation* rather than a *motion*. For the moment, let's think of the photon as a program that modifies the informational state of the destination D.

The source of information is the program that changes the informational state of the destination. The transformation takes a finite time, and since it is discrete, its effect is visible only after the transformation is complete. The result of a transform is a change in state, at the receiver. This means that the change in state must occur once the transform is complete and at that point we can say that the photon was 'received' at the destination. This would mean that a photon never travels from source to destination. The photon only causes a transformation to the state of a destination object. The destination object must be chosen as an input to the program before the program can begin executing. The 'goal' of the photon is therefore an input to a program, which must be chosen before the program begins executing.

The problem of the constant speed of light now reduces to the problem of how an input to a program is chosen.

Recall that in quantum theory energy cannot be emitted or absorbed unless a *choice* is made. Quantum theory predicts the effects of a changing force field but not whether that field will change. This missing causality in quantum theory can be bridged if the changing field is the effect of a choice. Furthermore, the constant speed of light can be understood semantically if the choice to emit energy is tied to the choice of the destination that must be modified by that emission. Rather than supposing that a photon is emitted, then travels to a destination, before it is absorbed by the destination, we could also say that the photon remotely modifies the destination. This would have been a problem in classical physics because non-local interactions were not allowed. It cannot be a problem in quantum theory where non-locality is already accepted. Also, we know that the destination of the particle must be determined *before* the emission starts. The particle cannot be emitted if the absorber is unknown. Since the 'entanglement' of the emitter and absorber prior to the emission is essential to an energy exchange, we cannot expect a local causality that happens after the exchange is complete.

The photon therefore does not travel in space. Indeed, it does not travel at all. It only acts on information to transform it.

There is, however, still a problem in identifying the destination object if the destination is moving. The problem arises because in quantum theory we cannot insist that objects at two different positions are the *same* object. Recall that objects at two different positions are semantically different states and we should ideally say that they are different particles. The particles that arrive at different positions in the double-slit experience, for instance, are indeed different particles. How then can we say that an object that changes its position *before* the photon 'arrives' at its destination is the *same* particle that was given as input to the program at the beginning of the transformation? If we cannot claim that the identity of an input to a program remains unchanged during the motion, then we also cannot apply the idea that the program acts on an input. This problem is illustrated in Figure-14 where the destination moves before the photon arrives at the destination. We can think of the photon as a 'long-running' program whose input changes before the program has completed. How can the program maintain its 'goal' if the goal is transforming? To sustainably speak about a program, we need the idea of a particle that does not change through its state changes. Is there something that does not change in state changes?

Source   Photon   Moving Destination

Figure-14 Moving Object Changes the Program Input

Electronic computers forbid simultaneous operations on the same symbol to prevent unpredictable outcomes. Mathematics does not permit two simultaneous operations on a state in parallel; mathematics only permits one causal action at a time. But this prevention is not possible in the real world where many causes can act simultaneously to change states. If a program takes a finite time to complete and the input's state changes in the interim, how can we consistently apply the operations of the program to its input? The finite speed of light tells us that the effect of a photon is 'added' to the other effects linearly, as if the effects of each photon were independent of the effects of other photons. This in turn implies that the effect of one cause does not 'interfere' with the effects of other causes. How do we conceive this causality if the state is changing?

This problem can be solved if we can re-induct the notion of an object back into quantum theory. Classical physics used both objects and states and it was possible to attribute many different states to the same object. The object-state distinction does not exist in quantum theory because a state is an object. If there was a notion of an object different from state in quantum theory, it would resolve the problem. Now, we could say that a photon acts on the *object* rather than the *state*. The problem would not arise if quantum theory had a notion about objects and causes acted on *objects* instead of *states*.

To introduce notions of objectivity in quantum theory we need new insights. One such insight is the role of the individual, separate from the universal. The state can be identified with the universal, which means the same individual undergoes a change in meaning.

However, the individual is the object that remains unchanged during state changes. Furthermore, the cause is directed toward the individual rather than toward the state. However, the changes occur to the state rather than the individual. We can rephrase this idea by saying that the time taken for a photon to be absorbed is spent not in traveling but in the absorption followed by the state change. Classically, we believe that the time is spent in traveling and the state change takes no time. Quantum mechanically, however, the idea of travel itself is problematic, and we cannot explain the constant speed of light based on quantum theory unless we say that the photon is directed toward the particle, rather than toward the state, and the time taken in causing a change is the time taken to absorb the photon rather than the classical time supposedly spent in a travel.

We have already noted that the individual and the universal are different spaces. So, when the state is changing, the change is occurring in the universal space alone. There is no change or movement in the individual space. Therefore, even if the state seems to change, the destination individual is fixed. Indeed, both the 'moving' photon and the absorber are individually static. The motion is simply to their states, rather than to their individuality. Therefore, if causality acted from one individual to another, there would be no problem of 'relative motion' and hence the problem of speed of light. Classically, we equate the position of the particle with the particle. But in quantum theory, that position is a *state* rather than the particle. The particle is in a separate space and combines with the state. By applying classical intuitions to quantum theory, we suppose that the particle is moving when only its state is changing. If we let go of this classical assumption, then there is no conundrum.

Quantum Gravity

It is well-known that quantum non-locality and relativistic locality represent contradictory pictures of nature, and without their reconciliation we would not have a unified theory of nature. The above discussion illustrated how we can reconcile quantum theory and relativity by separating questions of the individual and its state. The proposed solution suggests that the speed of light is constant for both moving and non-moving observers because the individual is not moving. Motion is a phenomenon about the change of state of the individual, not of the individual. This is consistent with quantum theory, which only speaks about the changes in state, without claiming that the successive states constitute the *same* particle. In fact, as we have discussed before, these states are different particles. The individual is undergoing state changes, by accepting different states, and the succession of these states cannot be predicted in current quantum theory. The 'motion' in quantum theory pertains to the redistribution of matter-energy into different states, rather than the motion of a single individual through multiple states.

Nevertheless, we still have the notion of a particle moving through states, although that notion doesn't exist in quantum theory because the 'particles' we speak of in quantum theory are states. To reinstate the intuition that a particle moves through states we need to conceive of an individual separate from the states. As we have said before, the state is the universal and the particle is the individual. The two combine to create a particle with a state, but it is possible to distinguish the state from the individual, because a different individual can have the same state, and the same individual can have a different state. Therefore, we need separate 'spaces' to distinguish between the individual and the state. Causality now acts on the individual rather than the state and causes a change of state. The new state of the individual existed previously as a possibility but has now become real. Therefore, the state hasn't changed; the individual has changed the state. The inability in quantum theory to predict the next state of the individual is about the fact that the theory doesn't deal with the individual; it only deals with states and their changes.

Every time we speak of an ensemble, we are talking not about a collection of individual particles but about the total energy and matter in the ensemble. This energy and matter can be redistributed in different ways, such that the total number of particles can increase, decrease, or remain the same. For instance, you can divide a mass of 10 kg into any number of particles without violating the conservation of mass. These divisions will create different number of individuals, although physics will be incapable of dealing with the resulting uncertainty about the total number of particles, because it is meant to deal with the total energy and matter. This idea of matter redistribution exists in General Relativity (GR) too, where matter redistributions enter because objects with the same mass can be swapped without causing any observable changes, *if* you happen to look at the universe from the 'outside.' If however you happen to be one of the observers being swapped in this process, then it will obviously create a perceptible difference. Arguments about indeterminism in relativity were foreseen by Einstein himself and he delayed the publication of GR due to these reasons. The formal statement of indeterminism in GR is called the Hole Argument[1]. Eventually, however, the problem was resolved through what is now called the Point Coincidence Argument in which the *events* in the universe do not change even when matter is redistributed. As noted above, this is true only if you look at the universe from the 'outside.' If you happen to be an observer in the universe, there is still an empirical difference for *you*. A detailed analysis of the indeterminism in GR is out of scope here, but I have mentioned the problem here only to connect it with the indeterminism in quantum theory.

Both quantum theory and GR are incomplete in the same sense: they *logically* permit many matter redistributions, which are empirically observable but cannot be predicted causally because these matter redistributions are consistent with the same total energy and matter. That incompleteness leads us to a simple insight: knowing the total amount of energy and matter does not fix the various ways in which matter can be distributed—first as the states and second as the particles occupying those states. If these matter redistributions represent information, the total amount of matter underdetermines the meaning that can be encoded in that matter. Of course, this should not surprise us at all, since the same computer hard-drive can encode many different pictures or videos, or the same amount of paper and ink can be used to print different books. Furthermore, which of these objects is which observer is also not decided because in quantum theory states are not individuals.

The flaw in physics begins when we fix the particles and their states and then compute the forces between these particles and conclude that all changes must be determined due to these forces. We don't realize that states and particles are different kinds of entities, and they can both exist in a state of possibility—i.e. without a combination—and they would be unobservable in this condition. To become observable, the individual and the universal must combine and there must be a relation to a measurement instrument. So, the observation is the result of this tripartite combination, but because these are separate types of entities, it is possible to distinguish them separately. Therefore, even if we fix the beginning of the universe as some particles in a definite state, the association between the particle and the state is not permanent. The particle can change state, and the state can change particles. We consider the former type of change, but not the latter type of change in classical physics, because we don't treat the state as something independent of the particle.

Quantum theory is unique in this sense; it indicates the existence of states without indicating the existence of the individual. That is, there is no conception of the individual thing that goes through successive states. We just have the succession of states. Also, since the states are only a redistribution of matter and energy, why the universe exists in a distribution is also unknown given that the total matter and energy are consistent with many matter redistributions. Both quantum and relativistic gravity deal in matter distributions, and what follows from conservation of matter and energy, which are indeterministic. In quantum theory, the indeterminism is the basis selection depending on the experimental setup. In relativity the indeterminism is that matter can be any redistribution. Beyond this state indeterminism, both relativity and quantum theory have the indeterminism about which individual is in which state.

We must therefore conceive of three kinds of laws. First, a law that speaks about the current state of matter distribution; evolution here pertains to the changes in states. Second, a law that speaks about the association between the individual and the state; evolution here involves the 'motion' of the individual through successive states. Third, a law that speaks about the interaction between individuals and states; the same state change can be caused by different interactions. Since each of these is indeterministic at present, it is difficult to visualize the ultimate form of the complete theory, but we can state why this understanding is a move in the right direction.

Physicists state that the unification of quantum and relativity theories is the greatest outstanding problem. I believe that even if these theories were unified, they will still not be the final theory because they are both set with the same problems of state, individual, and relational indeterminism, and the combined theory will have the same issues. It is not about the unification of a 4th force (gravity) with the remaining three forces (electromagnetism, weak, and strong). We misstate the problem if we state it in this manner, because we assume that the present quantum and relativity theories are themselves free of indeterminism problems, complete in themselves. The correct problem statement is that there are two theories, which separately have the same three kinds of indeterminism. Overcoming the indeterminism is the problem, not the unification of forces.

Nevertheless, within the context of the unification of forces, we can conceive of a theory that unifies the ideas of state distribution as a function of time. It won't solve the problems of relational and individual indeterminism—i.e. which particle has which state and which particle interacts with which other particle to cause that state change. But it can resolve the issue of state distribution as a function of time. This unification must be based on a quantum theory of gravity—i.e. conceive of gravity as a quantum rather than a classical force. This version of gravity will have the properties of quantum theory. For instance, particles with mass will not interact continuously; these interactions must be mediated by the transfer of a quantum particle (which is sometimes called the graviton). Similarly, we cannot think about a fixed distribution of matter independent of the observer. We must rather think about the total energy and matter in an ensemble, which can be redistributed by the observational apparatus (e.g. the number of slits in quantum theory). Finally, continuity of space and time in current gravitational theory must be discarded in favor of discrete units of mass and energy, that occupy discrete and orthogonal states and transitions between these states are discontinuous. With all these changes to the classical conception of gravity, it is possible to think of a unification using the ideas of information that I have outlined previously. Gravity will be conceived of as a different kind of information, more fundamental than the other three quantum forces because the particles dealt with by the current quantum theory have mass. So, the new theory must have mass, although as a different kind of information expressed symbolically rather than as a different physical property.

Antiparticles

Einstein's relativity introduced the equivalence between matter and energy. When Paul Dirac first incorporated relativistic considerations into quantum theory, surprising new conclusions came to light: it was theoretically possible to create pairs of matter and antimatter particles out of energy, which could then annihilate each other yielding energy. The idea of antiparticles requires another fact about information that we haven't discussed so far.

While science deals with facts or things that are true, meaningful things can also be false. For instance, 'The Sun rises in the East' is true but the statement 'The Sun Rises in the West' is false although meaningful. All true and false statements can be represented as symbols. When opposite meanings are expressed as symbols, they must have properties opposite to each other. Antiparticles can be viewed as symbols that represent these opposite meanings. If the particles represent true facts, then antiparticles are false.

In current quantum field theory, which combines non-relativistic quantum theory with special relativity, antiparticles have a positive mass, although they can have negative charge, spin, color, etc. We can imagine that when quantum theory is combined with general relativity, it would be possible for antiparticles to have negative masses. An antiparticle with a negative mass can be thought of as a particle with a positive mass accelerating in the opposite direction. The equivalence of inertial and gravitational masses allows us to treat negative inertial masses as negative gravitational masses, which can then be substituted by a positive mass accelerating in the opposite direction because of the equivalence principle. Negative masses will thus arise due to the presence of accelerating frames of reference in general relativity because in current quantum field theory, special relativity only deals with uniform motion or rest.

Of course, we will have to postulate that negative and positive masses have a repelling force between them instead of an attractive force between positive masses. An antiparticle with a negative mass moving away from a particle with a positive mass represents a decreasing repelling force which can be replaced by an increasing attractive force between particles if we substitute the negative-mass antiparticle by a positive-mass particle moving in the opposite direction. We will also have to postulate that negative masses experience opposite inertial forces than the positive masses. That is, a negative mass feels inertial force in the direction of its motion.

Since negative masses will repel positive masses, positive and negative masses can form stable regions of particles and antiparticles in the universe, whose total energy remains zero. This is different from present quantum field theory where particles and antiparticles quickly annihilate each other because their opposite charges attract and bring particles and antiparticles together. These regions of particles and antiparticles cannot communicate with each other because particles and antiparticles will never have proximity due to the mutual repulsion between objects that have opposite masses.

Within what physics has observed so far, the true facts far outstrip the false claims, because we use matter that represents facts to generate quantum experiments. However, this only means that matter and antimatter cannot co-exist in the same location or near each other; more specifically, they cannot exist within the same ensemble. This does not preclude the possibility that particles and antiparticles may abound in separate parts of the universe, which is enabled by the fact that negative mass particles will accelerate away from each other while keeping the total energy constant. With negative mass, matter and antimatter can be created even with zero total energy because antiparticles would have a negative energy. The creation of the universe from zero energy can be a possible approach to the question of the universe's origin. In current theories, huge amounts of energy need to be concentrated in a small region of space, which creates the problems of singularity. If matter and anti-matter are created simultaneously and then separated in space, singularity problems are avoided while keeping the total energy zero.

When we attempt to know something unknown, we postulate—either the unknown is X or it is Y—where X and Y represent opposite alternatives. Then we test for X and Y through experiments. Quantum theory suggests that the universe can consist of opposite meanings, although these meanings must exist far from each other. Specifically, these meanings are different ensembles, related through a semantic opposition. The fact that they exist in opposite sides of the universe will represent the idea that they are logically opposite. The annihilation of particle-antiparticle pairs means that contradictory claims cannot co-exist and if some particle-antiparticle pairs are annihilated, the opposition between different parts of the universe is reduced and they would be drawn closer. This drawing closer creates the possibility of more particle-antiparticle pairs to be annihilated, thus resulting in a rapid collapse of the universe into nothing.

If there is only matter in the universe, the universe prior to origin must have an infinite amount of energy. But, if the universe is treated as information, then particles and anti-particles represent a contradiction, which implies net zero information. An informational universe can be created from zero energy by injecting contradictions, which fly away to opposite sides of the universe, if space is semantic[2]. If time is semantic, these opposite meanings will exist at opposite duration points within a temporal cycle. This gives us some crucial insights into how matter and antimatter can transform into each other—essentially, it involves a time inversion. In classical physics, time inversion is conceptually very difficult because it seems incompatible with the experience of things moving forward. In cyclic time, however, there can be a forward movement accompanied by a time parity inversion; temporal points on opposite sides of a time cycle must 'move' in opposite directions, even though for on observer situated on a time cycle, time always seems to move forward.

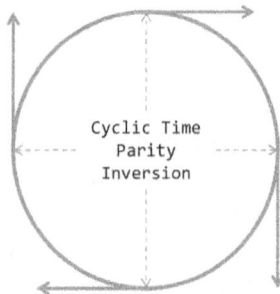

```
Cyclic Time
   Parity
 Inversion
```

Figure-15 Temporal Parity Inversion

A semantic understanding of antiparticles requires the hypothesis of cyclic time, which is linear and one-dimensional if we are 'riding' that time, but it inverts the parity leading to an automatic conversion of particles into anti-particles. Of course, observers riding on that time will not be able to physically measure the parity difference if they only observe their part of the universe at any instance of time. Even if they could somehow compare the particles in opposite sides of the universe (which is observationally forbidden since particles and antiparticles cannot communicate) they will still only see the opposite parity. The only way to know the parity inversion is if we can compare the past, present and future.

The everyday counterpart of this idea is the periodic rise and decline of ideas, cultures, societies and civilizations. The forward movement of time may not change our vocabularies, but it changes what we consider *true* in different times. The history of philosophy has, for instance, regarded empiricism and rationalism true at different times. Social philosophers have advocated individual good and collective good at different times. Civilizations have regarded matter and mind as fundamental at different times. The elementary concepts—e.g., empiricism, rationalism, individualism, collectivism, realism and idealism—have remained unchanged over time, but the truth we accord to these ideas has changed. A system evolves to a new ideology over time destroying other ideologies, only to return to the old ideologies at a different time. Societies with different ideologies cannot co-exist, although they can exist far apart. Typically, conversations between such ideologically opposed societies are difficult if not impossible. We might say that they are living in 'medieval times' as opposed to our 'modern times', signaling the fact that our culture thought in a similar way in the past.

Antiparticles represent all the everyday intuitive facts about ideological, social, cultural and civilizational evolution in time in the context of fundamental particles. Antiparticles can be understood as a parity inversion in time, which inverts the truth although not the meanings. Just as civilizations with different beliefs may exist apart from each other, it is possible to imagine that stable regions of the universe may abound exclusively in particles or antiparticles.

A particle and antiparticle pair represent two symbols at opposite ends of a time cycle. They are not therefore in the 'same' time, but if we observe them together we will find that they denote opposite meanings and hence truths. Antiparticles are not materially differ-ent 'stuff' as current quantum theory suggests by postulating nega-tive properties. Antiparticles can also be understood as particles at opposite points in a time cycle. In other words, antiparticles will not forever be antiparticles. With the passage of time, particles convert to antiparticles and vice versa, although their opposition to what were earlier particles remains intact as those particles are now antiparti-cles. Time inversion is therefore not an absolute property of a particle independent of other particles. It is rather the collective evolution of particle and antiparticle pairs in which time parity inverts periodi-cally and two particles always remain semantically opposite, although changing their meanings.

The Meaning of Spin

Quantum spin is a problem for interpretations because unlike other quantum concepts like position and momentum, spin does not have a classical counterpart. Even the everyday analogue of spin assumes an extended object, whereas classical objects are not actually extended. Quantum objects are extended, and the notion of spin should not be problematic. The conceptual difficulty in understanding spin, how-ever, arises from the fact that classical energy, momentum and angu-lar momentum correspond to the homogeneity of time, homogeneity of space, and isotropicity of space, respectively, but spin does not have a corresponding space-time symmetry property. This gap in interpret-ing spin can be bridged if it is associated with the isotropicity of time. The novelty of spin in the case of quantum theory can now be viewed as the fact that classical physics does not have time isotropicity since time is single dimensional and we postulate a direction in classical physics.

Direction in time naturally arises if time is treated cyclically rather than linearly. In linear time, there is forward movement and to go back in time we must retrace our steps, thereby inverting the trajec-tories. But, in cyclical time, there can be backward movement without retracing the steps. In fact, if time is cyclical, then the same object will be retracing its steps without going back in time. Cyclical time is a common feature of oscillatory phenomena because the direction of the motion of objects is reversed during a time cycle. The hands of a clock, for instance, move towards the right when the clock indicates the 12th hour and they move towards the left when the clock indicates the 6th hour. All quantum objects are described as vibrations and oscillations, but without revising the notion of time. This stance has its origin in classical physics, which began by describing rectilinear motion and has been extended into quantum theory where there cannot be motion without a change in energy state. Although the quantum particle is in a stationary state, it has energy, momentum, angular momentum and spin. The best way to understand the quantum stationary state is to think of it as an oscillation. These oscillations cannot be broken down into smaller linear motions because space and time are themselves indivisible.

When we associate spin with the direction in time, we are postulating something that remains conserved within the ensemble provided the ensemble's energy remains unchanged. But we also said that, in an oscillation, the time direction is inverted. How can a property associated with time isotropicity remain conserved if time direction changes periodically? The way to understand this problem is to see that the time is simultaneously inverted for all the particles in the ensemble, not for a single particle at a time. If, therefore, there are two particles with opposite spin in the same ensemble, they will switch their spins periodically. A chemical atom with odd spin will therefore periodically change its spin and this can be viewed as the reason for chemical atoms entering chemical bonds where the spins are even because this creates a stable ensemble of particles.

When direction in time is combined with the evolution of meanings, it is possible to interpret this direction as commonsense ideas about progress and regress or prograde and retrograde. What is progress and what is regress? These ideas cannot be defined simply by the passing of time; they must rather be defined in semantic terms. For instance, are ideas at the present progress over the ideas in the past? Ordering the occurrence of ideas as events in time is different from ordering them as progress or regress of ideas. Semantic notions about which ideas are better and hence progressive will need to be defined to interpret notions of direction in time.

The Weak Force

One of the most puzzling aspects of the quantum field theory is the weak force because (a) it acts at very short distances, and (b) it is mediated by massive W and Z bosons. These two aspects are inter-related, and the short distances of the weak force are explained by the fact that the W and Z bosons have mass unlike photons or gluons. The weak force mediates radioactive decay, the beta decay in which a neutron decays into a proton and a W boson which in turn decays into an electron and an anti-neutrino. To understand the importance of short distance it helps to recall that in SI the distance between quantum objects denotes semantic difference. To say that two objects are nearby is to imply that they are semantically similar. As symbols we might treat them as *synonyms*—things that are physically distinct but semantically similar. This semantic similarity creates a type of *internal symmetry* in the sense that these objects can be interchanged without causing observable differences. However, the property of all fermions is that they must exist as logically, and orthogonally, distinct entities and they could never be synonyms. The proximity of two quanta creates a physical condition in which the basic orthogonal property of eigenstates is violated, which would be unacceptable if these objects were to exist as fermions. This can be viewed as the root of the physical phenomena of particle decay.

The weak force must act at very short distances because only at such distances can two quanta become nearly synonymous to create an internal symmetry. As the distance grows, the particles become different and orthogonal. While the rest of quantum theory is based on logical distinctions, the weak force presents us with a case where two particles can get so close together as to violate the semantic difference between distinct quantum objects. Two quanta can never be in the same semantic state, and the particles must either be repelled farther apart or must be forced to decay into other particles. Decay is rare as compared to the more common phenomena of particles being repelled, but particles would be repelled only if they have the same charge. Both gravity and strong force are attractive and particles that happen to get close enough must decay.

The weak force can be viewed as mediating a decay of particles caused by the fact that the particles are close enough to substitute each other. This exchange symmetry implies that these particles are nearly identical in all respects and quantum theory forbids identical particles. The weak force is therefore unique and different from other forces such as gravity and electromagnetism that act at long distances and whose force particles have zero rest mass. The weak force requires redistribution of matter into new particles and matter needs to be transferred from one quantum object to another such that the new particles created from redistribution can be sufficiently different. The weak force bosons therefore carry mass, and this has been a peculiar aspect of quantum field theory because the bosons in other cases are massless. The weak force should not be viewed as a force but as a property of quantum objects where these objects cannot exist very close and if similar types of quantum objects are brought very close they will decay into dissimilar objects.

6

Comparing Interpretations

I am now convinced that theoretical physics is actually philosophy. It has revolutionized fundamental concepts, e.g., abut space and time (relativity), about causality (quantum theory), and about substance and matter (atomistics). It has taught us new methods of thinking (complementarity), which are applicable far beyond physics.

— *Max Born*

Introduction

This chapter compares SI with other prominent interpretations of quantum theory. Obviously, due to space constraints it is very hard to compare SI with every known interpretation here, and thus I have only chosen the following views to contrast SI against:

- the Copenhagen Interpretation

- the Statistical or Ensemble Interpretation

- the Many Worlds Interpretation

- the Von Neumann/Wigner Interpretation

- the Relational Interpretation

- the Objective Collapse Theory

A contrast with these well-known views serves to clarify the ideas behind SI, as well as to enable an understanding of how SI goes beyond clarifying quantum theory and into novel physical principles. In the following paragraphs, I will first briefly describe the key ideas underlying other interpretations before contrasting them with SI.

The Copenhagen Interpretation

The Copenhagen Interpretation (CI) derives its name from the work that Niels Bohr and Werner Heisenberg did in Copenhagen during the early days of quantum theory. CI interprets quantum theory literally according to the mathematics in the present form of the theory. It claims that the theory does not deal with *reality* but with observations and the theory tells us that we can only describe these observations probabilistically. The wavefunction ψ describes all possible observational states, whose probabilities can be derived according to Born's rule. Since ψ represents probabilistic states, it is not possible to know ψ as an object. At the point of observation, ψ collapses into one of its states but the collapse cannot be explained in current quantum theory. The properties of each quantum can't be measured at once—Bohr also maintained that they are not *defined* simultaneously—due to the uncertainty principle. Hence, we cannot draw trajectories of quantum objects in the classical sense. However, since quantum theory is primarily a theory of the sub-atomic reality, its effects are negligible at the macroscopic or classical level. Quantum theory will therefore approximate or reduce to classical physics at the macroscopic level. The real concepts that will describe quantum objects can't be known at the macroscopic level since we can only conduct experiments with macroscopic objects, and in quantum theory we have reached the limit to applying those classical physical concepts to the atomic and sub-atomic realities.

The main difference between SI and CI is the treatment of individual quanta. In SI, quantum theory is about the individual quanta and their dynamical properties can be given a non-classical interpretation. These properties are physical and observable, but they don't behave like classical properties. They can thus be treated as non-classical properties of quanta and given an informational interpretation. All other features of quantum theory—uncertainty, non-locality, and statistics—can be explained by postulating that the quantum dynamical properties are non-classical properties.

SI further claims that the notion of quanta must be applied not just to matter and energy, but also to space. Indeed, the *form* of objects is taken from the forms that space can take, and space is not infinitely divisible but can only be quantized into some fundamental shapes and sizes. Once we take away the infinite divisibility of space and time, we take away the ideas of continuity and the present mathematical definition of differentials, which depend on continuity. Position and momentum are thus never defined at once because (a) objects are always extended and hence don't have a fixed position and (b) for an extended object the rate of change of position with respect to time is not defined, because a particle's position is fixed for a small duration of time and Δx and Δt can't be arbitrarily small.

Quantum theory is statistical because we measure symbols by their physical properties and not according to the information content in them. Thus, we can't explain why physically identical electrons arrive at different locations in the 2-slit experiment. Distinct symbols will *always* arrive at different positions and an electron arrives at a unique location because it is a distinct symbol. By distinguishing electrons based on their informational properties, we will not get rid of probability, but we can explain it better.

Once we describe quanta as informational symbols, we can also explain non-locality because each unit of information is defined in contrast to an opposite meaning. Correlated pairs of particles with opposite spin are therefore pairs of opposite concepts like hot-cold. We might not presently know which particle has which spin (and hence which concept it represents) but the particle itself represents a unique and objective information content. As long as the particles are entangled, their conceptual distance remains unchanged, regardless of the physical distance. This physical distance, as we have noted earlier, is a function of the strength of the interaction. When we pull particles apart, each particle interacts with a different set of objects. But the entanglement is not to be described by this distance. It must be described by the conceptual distance, which remains unchanged even as the physical distance is being altered.

Thus, when a measurement is performed, non-local effects don't need to travel. Rather, the measurement always reveals the objective content in the symbol. One way to empirically prove this fact would be to note that particles with positive spin will always be detected at the same detector and particles with negative spin always at the opposite detector, regardless of the order in which we measure them. This experiment will show that the spin of a particle is not decided by the measurement but is *a priori* fixed by the particle's state itself.

SI and CI differ in the treatment of the wavefunction. In SI, the wavefunction represents the states of an *ensemble* of particles each of which has a definite state, but we currently don't know which particle has which state. In CI, different detections pertain to different *states* of the *same* quantum object and quantum theory is uncertain about the exact state of this object prior to measurement. The indeterminism in CI is that the individual quanta are in a superposed state and can never be resolved into a definite state without measurement. Indeterminism in SI is because we don't know which quanta is which symbol, but this can be resolved by looking at the *form* of the quanta and thus the indeterminism can be overcome.

If we apply CI to the analysis of letters in English, we would derive the counterintuitive view that sentences in a book aren't well formed in advance. Only when we read the book, some squiggle becomes the letter 'A', another squiggle becomes letters 'B', 'C', 'D', etc. We put together the squiggle sequence and make sentences out of it. This interpretation of quantum theory has two kinds of issues, only one of which is generally given importance. The first, and more prominent, issue concerns the fact that CI implies that there is no reality unless you measure it. Reality is defined at the point of measurement and so what you know is not what existed prior to being known. In so far as science is the discovery of things as they exist prior to being known, this is a direct attack on objectivity in science. The second issue (which hasn't been considered so far) arises when we apply quantum theory to books. As per CI, there is randomness in which letters would be observed upon measurement, so it is a miracle that we are able to ever read meaningful books. Of course, CI will say that books are macroscopic objects and hence quantum theory doesn't apply to them. But this means that we could not write books with atomic objects. That has profound implications for technology to be built with quantum objects where we could for instance use quantum objects to represent computer programs.

A sequence of 100 letters in the English language can be arranged in 26^{100} ways, only very few of which would be meaningful. Given the restrictions of what can be meaningfully said based upon what has already been said in prior chapters, paragraphs and sentences, even fewer sentences are meaningfully allowed. To narrow down many possibilities into a small number of actualities is the job of the author who chooses words and expressions to form letter sequences. This choice is in turn driven by an understanding of the subject. The sequence of letters in a book is not random. It is rather consciously crafted through a pre-arranged understanding of meaning and conventions through which these are expressed. Per the CI, however, the letters in a book are not pre-arranged. The identity of each letter is defined at the point of measurement, when a squiggle-wavefunction collapses into a definite letter. Since this collapse is random, the emergence of a letter out of a squiggle is also random. If CI is true, then a meaningless squiggle sequence becomes a meaningful sentence quite by chance. Since the probability of one meaningful sentence out of 26^{100} possibilities is extremely small, it is a miracle that we can encode and decode any information.

The problem of realism in CI is just a philosophical problem. But the problem of meaning is a scientific problem. By claiming that a wavefunction is in a superposed state, we introduce unavoidable indeterminism in quantum theory. SI does not have this problem. While we cannot predict the order of letters in a sentence today, we can say that they are distinct letters to begin with. The wavefunction is not about the superposed state of an electron. It is rather about the relative probability of finding one *type* of electron. It means that there must be ways to distinguish electrons that arrive at different positions (based on information), although current quantum theory has not formalized these methods. The different positions of an electron in the 2-slit measurement do not indicate randomness. They are rather combinations of two types of order. First, there is the order about the type of particle being observed. Second, there is the order of the type of relation in which that particle is observed.

This idea can be illustrated through an everyday example. We all have different abilities which exist in us as potentialities. For instance, we can eat, walk, talk, etc. However, these abilities are also exercised in relation to different external objects. For instance, we can eat different kinds of food. The observation involves a combination of these two things—an ability being exercised, and an object in relation to which this ability is exercised. These two things are present in the quantum 2-slit experiment as well. The particle being observed is like the ability in the measured system, but this potential can be exercised in relation to different measuring devices. The observed randomness can be attributed to the fact that we are unable to predict which ability is used in which order, and then which ability acts in relation to which measuring instrument. To explain the order, we need to predict both orders before we combine them.

Thus, the same *type* of particle can arrive at a different detector, and the same detector can receive a different type of particle. The eigenfunction describes a combination of these two possibilities. Both are types, but of two distinct kinds. As we have noted earlier, we need to think of three kinds of spaces—universal, individual, and relational. The arrival of a particle at a detector is the combination of a type of particle (universal) and the choice of a detector (relation). The conversion of a potentiality of universals determines the type of particles, and the conversion of the potentiality of relations decides the type of detector. The combination of the two produces the apparent randomness in the 2-slit experiment. The position of the detector only measures the relation, but the type of particle is undecided by this measurement, which is why the quantum measurement is bereft of the meaning of the detection.

The 2-slit experiment is like many observers (detectors) measuring a book. The book is a potential which manifests a symbol in an order determined by the causal nature of time. Similarly, the detector is chosen among one of the many possible detectors by the causal nature of time. This is akin to a person exercising their abilities in relation to different objects—we are required to select an ability and then select the target application of that ability. If we don't separately account for these two potentialities, then the eigenstates will be a multiplication of potential abilities and targets.

In classical physics, the *same* object can be at any position in space.

This is because of two reasons: (a) the identity of an object is independent of its location in space and (b) space is itself homogenous and isotropic. In quantum theory, this assumption must be modified. Every ability cannot be applied to every possible object. For instance, you cannot eat a table. However, you can still eat many types of food. Therefore, every particle cannot arrive at every detector, but there are many detectors that a particle can arrive at. This selective application of abilities restricts the multiplication problem mentioned above— we cannot linearly multiply all the abilities by all the objects to which the ability can be applied. There must rather be a selection of abilities based on their semantics and the target objects (again based on their semantics). A semantic interpretation of the particles and detectors is necessary to solve the measurement problem. Furthermore, an active role of time is necessary to decide which ability and relation manifests when.

The Statistical or Ensemble Interpretation

The ensemble interpretation (EI) given by Einstein is closer to SI in the sense that EI identifies the root problem in quantum theory to be the inability to describe the individual quantum object. Einstein took this position and viewed current quantum theory as an incomplete theory because the goal of physics according to him is the description of the individual quantum object. In SI, the description of the individual quantum object is completed by interpreting quantum dynamical properties not as classical properties but as meanings. This gives us new ways to explain quantum probabilities.

Einstein attributed the probabilities to the fact that quantum theory describes an ensemble and hoped that one day quantum theory will describe the individual quantum. This idea foregoes the fact that an ensemble is necessary for the quantum properties to be created. A single quantum cannot be a symbol of meaning because meanings always arise in a collection of quantum objects. It is therefore incorrect to assume that if we did describe the individual quantum then we would solve the quantum problem. Quantum theory is inescapably about the ensemble although the quantum can be understood as a symbol in the ensemble. As a symbol, the quantum is an individual, but its properties depend on the ensemble.

EI doesn't explain uncertainty and Einstein believed that classical concepts would be *en masse* replaced by new concepts in quantum theory. This is one of the prime reasons why Einstein and Bohr clashed on quantum theory because Bohr believed that classical concepts arise from our everyday intuitions and we can't replace them because to do experiments we need those intuitions.

In SI, we retain classical concepts like position and momentum, but discard infinite divisibility of space and time, which is inconsistent with everyday intuitions, and was added only to make calculus logically rigorous. If space is not infinitely divisible, then a chair is not located at a definite position as the chair is an extended object. Similarly, if time is not infinitely divisible then a quantum object *hops* from one location to another and the number $\Delta x/\Delta t$ is infinite at the instant it hops. The uncertainty of quanta can be explained by supposing that quanta are intrinsically extended.

EI doesn't touch non-locality and Einstein was a co-author on the EPR paper, although he later said that the problem could have been stated differently. The EPR paradox arises by applying quantum statistics, although treated in the toss-of-a-coin sense to a pair of correlated particles. We suppose that each quantum tosses a coin to decide the state in which it would be going next. But this is inconsistent with the ensemble view where a pair of correlated particles is in fact an ensemble of two objects and by definition the objects must be *a priori* distinct for the ensemble to be formed. Here, Einstein appears not to have taken his ensemble idea all the way but appears to have succumbed to the classical statistics problem.

One criticism of EI is that it does not adequately describe single systems or small ensembles because if we apply EI to a single particle, then that particle must be in a definite state, in violation of quantum theory. This criticism is presumptive because it assumes that quantum theory can be applied to isolated particles. My contention is that it cannot be because material concepts always require an opposition or distinction to exist, and hence can't be defined in isolation. EI can however be applied to twin-particle systems as shown earlier in the EPR paradox, although it requires us to treat statistics as draw-of-a-ball rather than toss-of-a-coin. The singleton criticism in facts point to a funda-mental difference between classical and quantum physics because all quantum phenomena must arise in collections. That is, a system of particles is partitioned according to logically orthogonal concepts but that requires an ensemble. Contrast this with Newton's physics, which began by defining a single, isolated particle before it defined the inter-action between particles.

EI falters in the case of Schrödinger's cat experiment where either the cat is dead or alive, and there is only one cat to consider. Accord-ing to EI, quantum theory describes an ensemble of many cat experi-ments set up one after the other. Observed probabilities now pertain to the fact that sometimes the cat is observed dead or alive depending on which instance of the experiment we are looking at. This is again a consequence of using classical statistical concepts in quantum the-ory, because observing results across many cat experiments is just like tossing the coin up in the air many times.

The cat paradox is related to how quantum theory is applied to time, and there is presently no quantum theory of time. The cat dies at the point of the radioactive decay. The cat is never in a superposed state; however, due to the radioactive decay the state of the whole system changes. The assumption that the cat is in a superposed state would never explain why, after the cat dies, we never see it alive in a quan-tum experiment (according to the theory, the probabilities remain the same before and after the cat is dead). To explain the fact that after the cat has died we never see it alive anymore, we must suppose that there is something that changes the wavefunction at death such that it has only one state after the cat's death.

The Many Worlds Interpretation

The many-worlds interpretation (MWI) initially proposed by Hugh Everett contends that every quantum interaction event causes the wavefunction to split into different 'worlds.' One of those worlds is the one that we currently see. But there are other worlds in which other possibilities in the wavefunction are observed. The splitting of the wavefunction into many worlds coincides with measurements.

MWI avoids the collapse thesis by assuming that the wavefunction never collapses into a single outcome. Rather, it collapses into all the outcomes simultaneously, but each outcome is a different world. The 'universe' is however different from each of the worlds. The universe is the objective wavefunction and each world is created from the splitting of this wavefunction. The worlds therefore are identical in the distant past but begin to differ radically as time progresses.

There is no way to verify if MWI is true because simultaneously being in different worlds to observe mutually exclusive alternatives is impossible. MWI however avoids (a) the quantum-classical boundary, and (b) the interaction with an observer to affect a collapse, which are two nagging issues that have troubled CI. In fact, in MWI, measurement has no special role as the wavefunction splits on any interaction, not specifically the measurement event.

There is little in common between SI and MWI because for MWI the wavefunction is objectively real and in SI the wavefunction is a function describing the probabilities of individual eigenfunctions; the eigenfunctions are real. Furthermore, all the eigenfunctions are real in SI within the same universe, while the eigenfunctions are real in MWI in different universes. In MWI, only the measurement outcomes are real at the point of measurement in the universe that measures. MWI does not explain uncertainty, non-locality or probability while these are explained based on the notion of meaning in SI.

MWI and SI also widely differ in terms of their overall goals. The main goal of MWI is to explain the measurement problem while keeping the mathematical formalism of quantum theory unchanged. SI, on the other hand, shows that the quantum theory formalism is incomplete because it doesn't explain the reasons that separate a quantum system from a classical one. In MWI, the everyday world is classical and in SI the everyday world is semantic. In SI, the main physical reason that distinguishes classical and quantum systems is the meaning in objects which arises in ensembles. Due to this meaning, a quantum object has properties which will not exist in the object outside the ensemble. These properties can denote concepts and names by which a quantum can describe other quanta. Classical physics is inadequate because it describes *independent* particles. Quantum theory under SI describes individual particles but they are not independent. Rather, a new notion of contextual individuality needs to be constructed. This individuation denotes meanings through the very same dynamical variables that current quantum theory views as the quantum extension of classical properties.

In SI, all quantum problems are apparent symptoms of the fact that the current theory describes information-carrying symbols as classical particles. A symbol can be interpreted as a thing, as a concept, or as a formula. The current theory however interprets a symbolic reality as a thing (particle), neglecting its informational properties. Probability results from the fact that a symbol is not fully described in the thing-like particle view. The wave-particle duality indicates that the thing-like quantum description must be complemented by a wavefunction of probable alternatives because the thing-like view is itself incomplete. The wavefunction however does not complete the particle view, leaving us with an incomplete and probabilistic theory. To solve this problem, the wavefunction must be replaced by the informational properties of the symbol.

Quantum uncertainty indicates that symbols are not point particles, because no symbol can be of size zero. To represent a finite amount of information, a symbol must have a finite size because through that form a thing becomes a symbol. Zero size means all objects are identical and their differences in position can be measured as a physical property but cannot be viewed as a semantic type distinction. All objects with information must therefore have a finite size. The smallest size symbols pertain to the smallest bit of information or the minutest operation that nature can represent. Planck's constant thus has a real meaning in terms of the smallest information in nature. Non-locality is a feature of information where symbols are entangled because their meaning is only given in relation to other objects in the ensemble. The entanglement is a non-classical property, but it does not obviate objectivity. As noted above, quantum objects are *individuals* but not *independent*. The notion of a classical object that must be independent to be an individual needs a revision in quantum theory, to be replaced by an individual that is contextual. Information must be represented through *distinctions* which involve multiple related objects, thus creating non-locality.

The Von Neumann / Wigner Interpretation

This interpretation was first proposed by John von Neumann, then expanded by Eugene Wigner but later discarded by Wigner. The interpretation claims that the wavefunction collapse is caused by an observer's consciousness that reduces the possibilities into one. Had it not been for the stellar reputation that von Neumann had already established for himself as a mathematician, the introduction of non-physical consciousness within physics would not have been kindly received. This interpretation is interesting for two reasons: (a) the extent to which physicists are prepared to go to solve the problem of measurement and (b) the realization that quantum theory has something to do with the mind. Von Neumann joined these two ideas into the claim that consciousness creates the observed world.

There are several difficulties with von Neumann's approach to the collapse. First, experimental evidence[1] has shown that a decision about the nature of reality (what we call a collapse) is made much before it hits an observer's awareness. In fact, in most cases, the collapse is effected automatically and unconsciously even before the observer becomes aware that such a decision has been made. Presuming therefore that the collapse is effected by an observer's consciousness is false from what we know from brain studies. Second, even if consciousness were to make the collapse decision the matter-consciousness interaction problem has proven notoriously hard to solve. We are far from explaining the nature of subjective processes like thinking, willing, emotions, etc., and the mind-body interaction problem is even harder to explain in comparison.

SI has a direct role for meanings and information, but no direct need for consciousness. Information can be objective since we can write books, compose music and paint pictures, but consciousness as self and identity is necessarily personal and subjective. I believe it is too early in physics to start talking about consciousness, because there are many things between the physical objective properties of classical physics and the consciousness of personal experience that still need to be explained. These include sensation, cognition, mental representation, intelligence, creativity, emotions, personality, the unconscious, ethical or moral judgments, and so on. Objective information is a logically next step in the study of material properties as they relate to observers, although information can be studied without explicitly invoking consciousness. If we are successful in understanding the nature of information, we might be able to extend this theory into an understanding of thinking processes and intelligence. Questions of consciousness lie beyond these questions.

SI uses subjective notions about meaning but objectifies them as properties of physical objects. This approach is needed to explain other aspects of quantum theory such as uncertainty, non-locality and prob-abilities, which seem to not involve consciousness at all. As we have seen, all aspects of quantum theory can be demystified by the simple premise that quantum objects are information-bearing symbols and their information content is depicted by their dynamical properties, which are defined in relation to other members of a quantum ensemble. This approach goes beyond current physics, and yet remains within the realm of experimental science. The measurement problem needs not an explanation of why the wavefunction collapses but which symbol in a sequence will turn up next. That, in turn, requires notions about closed space and cyclic time, using which we can show that the measurement of probabilities in quantum theory is like measuring a book in terms of word probabilities. These measurements aren't incorrect, but they miss the point that to know an informational object we should not be measuring word recurrence probabilities but observing the conceptual types symbolized by those words so that the resulting word sequences can be given meaning and predicted in a new way.

I do not believe that we are hitting a problem of consciousness in quantum theory yet. The problems that we are hitting are however related to conscious beings in the sense that meanings are understood and processed by living beings. Physics has encountered phenomena that are not as objective as classical physics to be totally divorced from observers but not entirely dependent on observers as to require consciousness within measurement procedures. Everything related to meanings can be studied within material objects as information, although the *insights* needed to develop this physics has to come from a better appreciation of the nature of minds. Success in physics will now require intuitions and insights about how the mind works, but those insights can be realized within matter.

The Relational Interpretation

The relational interpretation (RI) proposed by Carlo Rovelli claims that an observer-independent notion of reality in quantum theory is flawed because all descriptions of reality depend on some observer's perspective. The problem with CI, according to RI, is that it tries to construct an observer-independent account of the quantum state, presuming that the world is in a unique state. RI extends the relativistic idea that the result of observing depends on the observer's frame of reference to claim that quantum measurements are also different observer perspectives about the reality.

In the 2-slit experiment, for instance, each detector is a different observer that observes the same reality in a different way. Probability arises if an observer outside of these detectors tries to combine the detections of individual detectors, into a single observation. If, however, we treat those detections as distinct results obtained by different observers then there is no need for probability because there would be no need to combine those observations. RI says that different observations correspond to different *perspectives* of reality from different detector *viewpoints*. In quantum theory we cannot relate distinct observations to a single reality. Observations are therefore properties of relationships between the observer and the observed, and we can only speak of relations, not reality.

RI does not require a classical world and both the observer and the observed are quantum. But discarding the quest for reality is opposed to the original aims of science, which is to discover reality independent of observation. It is even more counterintuitive in the case of quantum theory because the theory predicts relative frequencies amongst observations and if these observations span multiple observers, then we must conclude that the theory makes predictions across many observers, of which there is no precedent.

It is possible to think of the detectors as different sense-organs of a single observer (like eyes, ears, skin and nose) in certain cases (it would be impossible to think this way for non-local measurements on objects that lie outside the light cone of any given observer, but in the 2-slit experiment we can think in this way). Taken this way, the relations between the detector and reality are in fact relations between reality and the sense organs, and they belong to one observer. They still involve relations but between various parts of a single observer, and relativity-style ideas can't be applied. But even when the measurements span across different observers as in the case of non-local phenomena, the issue is not that separate observers measure it, but that one observer can *know* (although not experience) the outcome of the second observer's measurements. The curious aspect of quantum theory is that it enables predictions outside an observer's domain of experience, which is something that we expect an interpretation to explain and not merely attribute it to the differences of relations across observers. The relational viewpoint explains why reality is seen differently across observers, but it doesn't explain the relationship between these viewpoints, unless of course we add rules of transformation across these observers, like the Lorentz transformations in the case of relativity. RI doesn't provide these transformation rules, and hence the interpretation doesn't explain probabilities or the reason for non-locality.

RI assumes that there is a single reality being observed by observers. That may not be so, if there is an ensemble of particles and each observer sees different things because there are in fact different things and each observer can see a part of the whole ensemble. The latter is how we think of reality in SI and each quantum arrives at a different detector because it is a different *type* of quanta and because different quanta arrive at different detectors. In SI, thus, different detectors obtain different results because there are multiple realities given by different eigenfunctions. SI can be regarded relational in the sense that a single relation picks out a certain *type* of quanta quite like how eyes will see color and nose will pick out smells. But to recognize this, we need to acknowledge that there are different types of realities which are picked up by observation via relations.

RI opposes hidden variables and SI does not add new variables. It rather reinterprets the eigenfunction form as a symbol and the dynamical properties of the eigenfunction as meanings of the symbol. Meanings in an ensemble are related but conceptually orthogonal. The detectors via interaction with the observed world acquire different states to detect different meanings. This is like how a detector becomes an eye to measure color and another one becomes a nose to measure smell. Both color and smell are simultaneously real and present in the same reality although a different type of observational relation is required to pick out smell versus color.

Probably the biggest difference between SI and RI is that SI explains why quantum theory is probabilistic. In SI, the probabilistic description of reality is like describing a book in terms of symbol probabilities. Knowing the symbol frequencies is partial knowledge. The meaning in a book arises both due to symbols and their order. Currently, that symbol order seems random because we attribute causality to the object rather than to time, which triggers the potentialities. The causality in time requires a hierarchical and cyclic view of time. Similarly, the understanding of matter needs a hierarchical and closed view of space. Both produce semantics, and their interaction produces an effect. Thus, the world is objectively possibilities, but which possibility will be converted into a fact depends on the possibilities and the time triggering them.

Since observation depends on a relation, and the relation has both a knower and a known, each of which exists as a possibility, it is important to account for both sides of the observation event. For example, as a possibility in an observer is triggered, through a relation to a known, the observer can see the same thing in the known. The cause of that knowledge would be the observer as it precedes the conversion of the possibility in the known, but the knower is also triggered by time. In that sense, ultimately, the causality lies in time and through intermediaries it can also be present in material objects. While it is possible to *explain* the effect based on intermediaries, it is not possible to *predict* those effects without time. Time can however both explain and predict.

Therefore, the problem with RI is that even if we attribute the observation to a unique observer or a relation to an observed, this is merely an explanation, not a prediction. Quantum theory will remain probabilistic without this predictability, and RI will remain just an interpretation of the current formalism, rather than something by which we can fix the root cause of the problem—i.e. predictions. RI can however explain why different observers can interpret the book differently depending on whether they see it or smell it, or the way their brains view the book such that the concepts in the world are

adapted to the concepts of thinking. This entanglement can recipro-
cally alter the concepts in the observer or the concepts in the external
world, using decoherence as the basic mechanism.

The Objective Collapse Interpretation

Objective collapse (OC) is not an interpretation but represents various
mechanisms that will either modify or complement the conventional
quantum theory formulation. The Ghirardi-Rimini-Weber theory
(GRW) for instance modifies the Schrödinger equation to include
non-linear factors which objectively collapse the wavefunction without any external influence. Roger Penrose similarly postulates that
gravity waves collapse the wavefunction.

In SI, the wavefunction never collapses because it is never superposed to begin with. The wavefunction describes not the *state* of an
individual quantum; in fact, the wavefunction doesn't describe states
at all. The wavefunction prescribes probabilities of finding certain
symbols whose typology is not understood within current quantum
theory because the theory treats space and time in a physical and
non-semantic manner. When space and time are treated in a hierarchical and nested manner, then it would be possible to interpret locations and directions in space and time as *types*. SI therefore indicates
a new mathematical formulation of quantum theory. What we think
is a collapse is a certain type of symbol being detected. A sequence of
symbols represents a description of information content which could
be pointing to other objects. Collapse and probabilities are therefore
emergent phenomena.

Conventionally, this view was adopted by hidden-variable theories
that explained collapse and probabilities by postulating additional
factors that existed but could not be observed. In objective collapse,
there are additional factors but they either change the equations of
motion or go outside quantum theory to affect the collapse. In SI, there
are no additional factors. SI reinterprets the dynamical variables in
current quantum theory as meanings. This helps us conceive how
quantum order amongst events—which is empirical but not explained
by current theory—can be explained. However, to form a predictive

theory of semantic phenomena requires a new space-time theory in which locations and directions in space-time denote types rather than quantities. By treating space-time as a domain of semantic information rather than as a domain of things, we can address all interpretive problems in the theory.

Epilogue

Quantum theory cannot be understood solely from its present mathematical formalism. To explain the import of the theory, different interpretations have added different ideas to the theory. The Copenhagen Interpretation adds a collapse; the Ensemble Interpretation adds statistics; the Many Worlds Interpretation introduces many possible worlds evolving in parallel; the von Neumann Interpretation added consciousness to the experiments, the Relational Interpretation adds many possible orthogonal perspectives and the Objective Collapse programs either modify the equations or add mechanisms to collapse the quantum wavefunction. SI also adds a new idea to quantum theory which is that space and time are domains of meaning and not domains of things. Symbols are also things, but things are not symbols. A theory of symbols can therefore extend the current formalism in quantum theory without adding new observables. A symbol acquires meaning in relation to other symbols within an ensemble while things exist independent of other things. A symbol is therefore an *individual*, but it is not *independent* of other things. The classical description of reality in which the world is independent things is flawed. Quantum theorists suppose that this means that the individuality in the classical sense is now lost. SI shows that it is possible to have individuality without independence. Since we have individuality, all the classical dynamical properties that we associated with independent objects can also be associated with the individual quantum objects. However, the dynamical variables must be viewed in a way to suggest individuality without independence. This is the way of semantic information.

Since classical dynamical variables are derived from properties of space-time, reinterpreting these variables entails that we reinterpret space-time. I showed this reinterpretation requires us to view locations and directions in space-time as denoting meanings or types. The current theories of space-time cannot be viewed in this way because in current science both space and time are infinitely extended and linear. A new theory of space-time is needed to view locations and directions semantically. In this view, space is closed, and time is cyclic. Like we describe postal addresses by nesting an object within a hierarchy of increasingly larger domains of space, or like we describe time as cycles of day and night, it is possible to describe space and time in a new way. When space is closed, and time is cyclic then directions in space and time are also described differently. For instance, we can speak of directions in terms of types such as East and West which can be subdivided into further refined types such as North-East or South-West. We can describe direction in time as the combination of the directions of various nested cycles; in each cycle, meaning is moving in a different direction and the combination of these directions will represent a trend in meaning.

With a new space-time theory, locations in space will denote concepts, locations in time will denote order, directions in space will denote oppositions in types, and directions in time will denote trends in meaning evolution. Momentum, energy, angular momentum and spin will represent program meanings corresponding to descriptive meanings denoted by locations and directions in space and time.

The question of the viability of an interpretation depends on the viability of the assumptions that we add for interpreting. In that regard, the assumptions added to SI are more viable as compared to assumptions about collapse, consciousness, many-worlds, statistics, complementary perspectives or modifications to quantum equations. Finite divisibility of space and time is certainly woven into the fabric of physics today, given the uncertainty relations. Hierarchical notions about space and time are common in the everyday world. SI attempts to combine the finite divisibility of space-time in modern physics with the typed-theory of space-time in the everyday world. My claim is that if space-time is semantic, then finite divisibility is itself the ability in space-time to denote meanings, quite like symbols can denote meanings. I believe this move paves a path not just for understanding quantum theory but also for how this highly successful theory can be integrated with our everyday view of reality and what that means in terms of newer kinds of experiments.

Classical physics began when Newton conceived an isolated particle defined by the fact that it moves in a straight line. In quantum theory, there are individual objects, but not isolated. When such objects
interact, the result is quantum entanglement. Entanglement creates
properties that an object did not have outside the ensemble. These can
be used to represent meanings. These meanings include sensations
or concepts. The experience of the meaning depends on the position
of the symbol in the hierarchy; abstract meanings cannot be sensed;
they can only be known. Contingent meanings can both be sensed and
known. Thus, quantum theory is the description of the world in the
ways that we perceive it as sound, taste, smell, sight and touch. Indeed,
now, dynamical properties of quantum theory will be measured as
physical properties but will *denote* concepts.

A quantum is a *phoneme*. As a vibration it is like ordinary sound
vibrations. This vibration can be studied as physical properties, if
we describe the vibration in an *individual* quantum. Like tones produced by musical instruments are not notes unless we collectively see
them as parts of a musical scale, similarly, these phonemes don't have
meanings unless they are viewed collectively with other phonemes.
Classical dynamical variables implied that an individual object is also
an independent object. Quantum dynamical variables imply that an
individual object is not independent. Therefore, while we can *measure*
the individual object's properties like we did in the case of classical
physics, we cannot treat these objects independently. It is now better
to think of particles as symbols whose physical properties are defined
collectively rather than independently. To formalize this idea, however, we would have to change our theories of space and time from
open and linear territories of material objects to hierarchical, closed,
and cyclic, as the domains of symbols.

There is a need today to better understand the behavior of symbolic
systems that include human and animal brains but can also include
semantic computers in the future. SI fulfills that need, as it describes
how atomic particles behave as symbols and how this behavior is different from everything we know through classical physics. If nature
is symbolic and matter carries information, then the universe is quite
different from the way we have thought of it so far. Space and time are
not a uniformity of points but comprise meaningful forms that can be

understood as *concepts* and *actions*.

Our everyday language makes use of concepts and actions to describe things and their changes. If nature itself defines a language of concepts and change using properties of space-time, then there is a natural basis underlying everyday semantics. By investigating that structure of space-time, we can uncover the concepts that underlie the everyday world. Personally, this possibility seems exciting to me, and I think it should be interesting to a lot of people across a wide range of concerns such as linguistics, anthropology and biology.

Immanuel Kant once thought that space and time are the basic categories or goggles through which we see the world. That idea was considered disproven with the rise of non-Euclidean geometries. But, with SI Kantian thinking has a more profound basis in the meanings which allow us to describe the world in terms of concepts. The manners in which we can give meanings to space and time become the manners in which we can think of the world as types. That in turn underlies everything that we can think of and express in language. The semantic properties of space and time can then become the basis of understanding meanings in language and in the mind!

References

Bell, J.S.: On the Einstein-Podolsky-Rosen paradox. Physics 1, p. 195-200 (1964).

Bohr, Niels: Atomic Theory and Description of Nature. Cambridge: Cambridge University Press (1934).

Culler, J.: Saussure. Glasgow: Fontana/Collins (1976).

Leach, Edmund: Lévi-Strauss. Glasgow: Fontana/Collins (1970).

Einstein, Albert in P. A. Schilpp, ed.: Albert Einstein: Philosopher-Scientist. New York: Harper & Row (1949).

Einstein, A, Podolsky, B. and Rosen, N.: Can quantum-mechanical description of physical reality be considered complete? Phys. Rev. 47 777 (1935).

Von Neumann, J.: Mathematical Foundations of Quantum Mechanics. Princeton: Princeton University Press (1955).

Zurek, W. H.: Decoherence, einselection, and the quantum origins of the classical. Reviews of Modern Physics, 75, 715 (2003).

Landau, L.D. and Lifshitz, E.M. (1980), Course in Theoretical Physics, (Pergamon, Oxford).

Penrose, Roger. (1989), The Emperors New Mind, (Oxford: Oxford University Press).

Jammer, Max (1966), The Conceptual Development of Quantum Mechanics, (McGraw-Hill).

Endnotes

1. QUANTUM INFORMATION

1   The eigenfunction is a mysterious concept. It is the "ghost" that holds information about all the quantum's properties and yet this "ghost" is itself not given reality in quantum theory. Mathematically, the eigenfunction is a complex valued function. Operators act on this eigenfunction to determine the outcome of measurements.

2   Bell, John (1964). "On the Einstein Podolsky Rosen Paradox". Physics 1 (3): 195–200.

3   The problem of perception also involves the interpretation of meanings from sensations. For instance, red can denote danger, passion, war and fear. A similar phenomenon is seen when the seven colors in the rainbow are used to represent the seven notes of music (in what is sometimes called synesthesia). This underscores the basic possibility that a sensation can convey more than what science measures from it. The dance of lights can represent a musical orchestra and the eyes cannot know that unless a semantic understanding of distinctions is applied to it (synesthetes know how to see both). In such cases, the sensation is color, but the meaning is sound. In other cases, sensations may be sound, but the meaning can be touch, taste or smell. There is hence a difference between sensation and concept, even when the same words are used to denote both. The difference is that sensations can be defined universally but concepts must be defined contextually through distinctions.

2. THE QUANTUM PROBLEM

4 Zeno was a Greek philosopher. He formulated many paradoxes which essentially show the impossibility of motion. One such paradox says that if an object must move from point A to B, then it must first go half-way through to B. But before it can go half-way to B, it must go one-fourth of the way, and so on ad infinitum. Zeno concluded that for an object to move from point A to B required an infinite number of steps and therefore motion was impossible.

5   Einstein, A; B Podolsky; N Rosen (1935-05-15). "Can Quantum-Mechanical Description of Physical Reality be Considered Complete?". Physical Review 47 (10): 777–780.

6   Penrose, Roger (1999) [1989], The Emperor's New Mind (New Preface (1999) ed.), Oxford, England: Oxford University Press, pp. 475–481.

3. DEVELOPING THE INTUITIONS

7   Stapp, Henry (2007), The Mindful Universe, The Frontiers Collection (Book 2), Publisher: Springer; 2nd ed. 2011 edition.

8   There is a basic difference between an order in time and a direction in time. Events can, for instance, be sequenced as in an order A, B, C without giving them an order. This order will be consistent even with time reversal which will convert the order A, B, C into an experience of C, B, A. This difference between order and direction was first described by McTaggart who called these notions of time tenseless and tensed. He called them A-series and B-series.

9   Quantum theory permits many eigenfunction bases in which the dynamical properties are different. The bases can be chose through a change in experimental setup. I will assume for the sake simplicity here that the experimental setup is fixed and therefore the dynamical properties are fixed too.

10 The term stationary state is a little misleading because all states are vibrating. However, the real and complex parts of the vibrations may be

vibrating in opposite directions such that if we computed the probability distribution using the Born's rule—i.e. $|\psi|^2$—it would remain constant in time. The probability distribution can also evolve in time like a vibration.

11 For a detailed discussion of this and other mathematical paradoxes, their root causes and possible solution, the reader is referred to my book *Gödel's Mistake*.

12 See note 10 for clarification on stationary states.

13 This claim is oversimplifying but not entirely inaccurate. I will later show that there are four different properties of space-time—locations and directions in space and locations and directions in time—which can be used to represent meanings. Generalizing these representations as a property of location in space suffices presently.

14 The 2-slit experiment comprises a quantum source that emits quantum objects. These objects are passed through two slits before they impinge on a battery of detectors behind the slits. The experiment shows that quantum objects don't arrive at all possible locations on the detectors. Rather, an interference pattern is generated such that quanta arrive at some locations and not at others. This interference pattern was explained in classical physics based on wave theories which were said to interfere. In quantum theory, it is observed that if the intensity of emission is reduced, then it is possible to observe individual quantum objects. This affirms the idea that quanta are indeed particles and not waves. However, to explain the interference pattern, a wave behavior is assumed, and the quantum state function is therefore called a 'wavefunction'.

15 See note 10 for clarification on stationary states.

4. THE SEMANTIC INTERPRETATION

16 Statistical mechanics arose from the need to describe the irreversibility of thermal phenomena using classical concepts. An example of such a phenomenon is the expansion and contraction of a gas in a chamber. If the gas is described by classical physics, which is a time-reversible theory,

then thermal phenomena should also be reversible. To explain the opposite observed facts, Maxwell and Boltzmann postulated that an ensemble of
classical particles is not in a definite state, but in one of the many possible
states *at once*. The total number of states was used to compute the entropy
of a system, which was then used to explain thermal phenomena.

http://link.aps.org/doi/10.1103/PhysRevLett.109.100404 and http://
www.nature.com/nphys/journal/v8/n3/abs/nphys2194.html.

17 Wigner, E. P. (1960). "The unreasonable effectiveness of mathematics
in the natural sciences. Richard Courant lecture in mathematical sciences
delivered at New York University, May 11, 1959". Communications on Pure
and Applied Mathematics 13: 1–14.

5. ADVANCED QUANTUM TOPICS

18 A point labeled by number M on N^{th} dimension can be represented by
the number PN^M on a single dimension where PN is the Nth prime number which is raised to the power of M to obtain a number for the single
dimension.

19 I use the word stationary state somewhat loosely in the sense that an
object remains in a state of oscillation. Some of these oscillation states will
indeed have constant dynamical observables (such as position) but other
states may indeed be observed to oscillate. A state of oscillation is still not
a state of motion in the sense that light travels from one point to another
and never comes back to its original position (which it would if it was in
an oscillatory state). While, classically speaking, light is also an oscillatory
electromagnetic wave, this is not the same as the quantum stationary states
where the objects indeed oscillate in a time-invariant manner.

20 Norton, John, 'The Hole Argument', The Stanford Encyclopedia of Philosophy (Spring 2008 Edition), Edward N. Zalta (ed.), URL = <http://plato.
stanford.edu/entries/spacetime-holearg/>.

21 Antiparticles with negative mass also represent an obvious explanation
for the observed expansion of the universe, which prevents the universe

from collapsing quickly. Note that not only is the universe expanding but the rate of expansion is increasing (the expansion is accelerating rather than decelerating). This fact can't be explained only by postulating a repulsive force between matter and antimatter, but also by supposing that the amount of matter and antimatter is increasing constantly. Since mass and negative mass can be created from a vacuum of zero energy, postulating an increase in the matter requires no additional physical principles. Instead, we suppose that more of the unknown in the universe is being decoded into opposite possibilities which create particle-antiparticle pairs.

6. COMPARING INTERPRETATIONS

22 Libet, Benjamin; Gleason, Curtis A.; Wright, Elwood W.; Pearl, Dennis K. (1983). "Time of Conscious Intention to Act in Relation to Onset of Cerebral Activity (Readiness-Potential) - The Unconscious Initiation of a Freely Voluntary Act". Brain 106: 623–642.

Index

9 7 8 8 1 9 3 0 5 2 3 7 2